AF659086

Ecological Assessment of FOOD CROPS

Ecological Assessment of FOOD CROPS

Editor

Y.S. Reddy

2006

Gene-Tech Books

New Delhi - 110 002

2006, © Publisher

Information contained in this work has been published by Gene-Tech Books and has been obtained by its author(s)/editor(s) from sources believed to be reliable and are correct to the best of their knowledge. However, the publisher and its author(s) make no representation of warranties with respect of accuracy or completeness of the contents of this book, and shall in no event be liable for any errors, omissions or damages arising out of use of this information and specifically disclaim any implied warranties or merchantability or fitness for any particular purpose.

All rights reserved. Including the right to translate or to reproduce this book or parts thereof except for brief quotations in critical reviews.

ISBN 81-89729-06-3

Published by : GENE-TECH BOOKS
4762-63/23, Ansari Road, Darya Ganj,
NEW DELHI - 110 002
Phone: 41562849
e-mail: genetechbooks@yahoo.co.in

Printed at : Tarun Offset Printers
Delhi

PRINTED IN INDIA

Preface

Natural ecosystems were disturbed on the day some 11000 years ago that the first planet dwellers decided to dig holes in the ground and plant seeds and tubers. Today the planet has to cope with over 6 billion people and urbanization, industrialization, abject poverty, while agricultural practices continue to impact on the environment. Assessing impact on the environment is much more than just a tunnel vision approach on hypothetical risks of genetically modified (GM) crops. Approval of genetic modification and new conventional varieties with potential risk should be subject to prior assessment for adverse impact on the environment. Identifying problems pro-actively is substantially better than corrective action after the problem has surfaced.

This book focuses on the ecological assessemet of genetically modified food crops. It covers in detail the development, use and regulation of GM crops. The applications and risks of food crops are also described in detail. The book is written for students, researchers and professionals in plant biotechnology and related sciences such as plant science, biotechnology, bioscience, environmental science and food science. It will stimulate readers thinking on key constraints in agriculture. Readers will get acquainted with a wide range of information, technologies and methodologies.

Editor

Contents

Contents

1

Ecological Assessment of Genetically Modified Crops

Genetic changes in plants, animals and micro-organisms have taken place with increasing impact and are as old as agricultural practices itself, as farmers domesticated and selected biodiversity. South Africa grows some 100 000 hectares of triticale, a wheat-rye cross, and consumers can enjoy plumcots, pluots and apriums from apricot-plum crosses, all coming from non-genetic modification techniques. Vegetative propagation through budding, grafting, tubers, rootstocks and tissue culture are major industries. Genetic modification (GM) application only started when scientists began to understand the periodic jump of genes from their positions on one chromosome to other positions in the genome (transposons), micro-organisms being able to incorporate foreign naked DNA (transformation) and vectors carrying genes from one species to another (transduction).

The first applications of GM technology were in human health (insulin in 1982), food processing (chymosin enzyme), industrial production, and, only by the early 1990s, GM crop plants. Genetic engineering is an application of biotechnology involving the manipulation of DNA and the transfer of gene components between species in order to encourage replication of desired traits. Although there are many applications of genetic engineering in agriculture, the current focus of biotechnology is on developing herbicide tolerant crops and on pest and disease resistant crops.

Transnational corporations such as Monsanto, DuPont, Norvartis, etc. which are the main proponents of biotechnology view transgenic crops as a way to reduce dependence on inputs such as pesticides and fertilisers. What is ironic is the fact that the biorevolution is being brought forward by the same interests that promoted the first wave of agrochemically-based agriculture, but this time, by equipping each crop with new "insecticidal genes," they are promising the world safer pesticides, reduction on chemically intensive farming and a more sustainable agriculture.

As long as transgenic crops follow closely the pesticide paradigm, such biotechnological products will do nothing but reinforce the pesticide treadmill in agroecosystems, thus legitimising the concerns that many scientists have expressed regarding the possible environmental risks of genetically engineered organisms. The most serious ecological risks posed by the commercial-scale use of transgenic crops are:

— The spread of transgenic crops threatens crop genetic diversity by simplifying cropping systems and promoting genetic erosion;

— The potential transfer of genes from pesticide resistant crops (HRCs) to wild or semidomesticated relatives thus creating super weeds;

— HRC volunteers become weeds in subsequent crops;

— Vector-mediated horizontal gene transfer and recombination to create new pathogenic bacteria;

— Vector recombination to generate new virulent strains of virus, especially in transgenic plants engineered for viral resistance with viral genes;

— Insect pests will quickly develop resistance to crops with Bt toxin;

— Massive use of Bt toxin in crops can unleash potential negative interactions affecting ecological processes and non-target organisms.

The above impacts of agricultural biotechnology are herein evaluated in the context of agroecological goals aimed at making agriculture more socially just, economically viable and ecologically sound. Such evaluation is timely given that worldwide, there have been over 1,500 approvals for field testing transgenic crops (the private sector has accounted for 87% of all field tests since 1987), despite the fact that in most countries stringent procedures are not in place to deal with environmental problems that may develop when engineered plants are released into the environment.

A main concern is that international pressures to gain markets and profits is resulting in companies releasing transgenic crops too fast, without proper consideration for the long-term impacts on people or the ecosystem. Most innovations in agricultural biotechnology are profit driven rather than need driven, therefore the thrust of the genetic engineering industry is not really to solve agricultural problems, but to create profitability. This statement is supported by the fact that at least 27 corporations have initiated herbicide-tolerant plant research, including the world's eight largest pesticide companies Bayer, Ciba-Geigy, ICI, Rhone-Poulenc, Dow/Elanco, Monsanto, Hoescht and DuPont, and virtually all seed companies, many of which have been acquired by chemical companies.

In the industrialised countries from 1986-1992, 57% of all field trials to test transgenic crops involved herbicide tolerance and 46% of applicants to the USDA for field testing were chemical companies. Crops currently targeted for genetically engineered tolerance to one or more herbicides includes: alfalfa, canola, cotton, corn, oats, petunia, potato, rice, sorghum, soybean, sugarbeet, sugar cane, sunflower, tobacco, tomato, wheat and others. It is clear that by creating crops resistant to its herbicides a company can expand markets for its patented chemicals. The market for HRCs has been estimated at more than $500 million by the year 2000.

Although some testing is being conducted by universities and advanced research organisations, the research agenda of such institutions is being increasingly influenced by the private sector in ways never seen in the past. 46% of biotechnology firms support biotechnology research at universities, while 33 of the 50 states have university-industry centers for the transfer of biotechnology. The challenge for such organisations will not only be to ensure that ecologically sound aspects of biotechnology are researched and developed (N fixing, drought tolerance, etc.), but to carefully monitor and control the provision of applied non-proprietary knowledge to the private sector so as to protect that such knowledge will continue in the public domain for the benefit of all society.

Biotechnology and Agrobiodiversity

Although biotechnology has the capacity to create a greater variety of commercial plants, the trends set forth by TNCs is to create broad international markets for a single product, thus creating the conditions for genetic uniformity in rural landscapes. In addition, patent protection and intellectual property rights spoused by GATT, inhibiting farmers from re-using, sharing and storing seeds raises the prospect that few varieties will dominate the seed market. Although a certain degree of crop uniformity may have certain economic advantages, it has two ecological drawbacks. First, history has shown that a huge area planted to a single cultivar is very vulnerable to a new, matching strain of a pathogen or pest. And, second, the widespread use of a single cultivar leads to a loss of genetic diversity.

Evidence from the Green Revolution leaves no doubt that the spread of modern varieties has been an important cause of genetic erosion, as massive government campaigns encouraged farmers to adopt MVs and to abandon many local varieties. The uniformity caused by increasing areas sown to a smaller number of varieties is a source of increased risk for farmers, as the varieties may be more vulnerable to disease and pest attack and most of them perform poorly in marginal environments. All the above effects are not ubiquitous to MVs and it is expected that, given their monogenic nature and fast acreage expansion, transgenic crops will only exacerbate such effects.

Firstly, we need to understand what "risk" is all about. The identification of risk starts with considering hypothetical scenarios based on past experience, i.e., the "what if" approach. Identifying a potential risk is next, followed by the probability that the risk will happen, and by analysing the potential consequences of such an event. It is good to remember that consequence do not equal damage. Next is developing risk management systems. Most risk approaches stop here and do not go to the next steps, namely, can regulatory systems be practical, implementable and affordable, or will it stifle progress and innovation? Consider the regulation of risk versus the risk of regulation.

The comparator must always be the baseline of what we already have (e.g., conventional crops), as well as the risk of not adopting a specific new technology. Approval by authorities of new technologies should be based on precaution. This presents an opportunity to make a decision in the absence of scientific certainty, and such a decision could be approval, delay until further information become available, or rejection.

In dealing with new plant species or new varieties, GM or conventional, risk analysis requires botanical information: dispersal of pollen, seeds or vegetative parts, and the frequency thereof. If the chance of these is zero, then the risks of invasiveness are zero. The nature and action of the gene has to be described. Without such information, sweeping statements like genes jumping into nature, horizontal gene flow and crosspollination, are meaningless. The typical concerns about potential adverse impact of GM crops on natural ecosystems can be considered as follows:

Will GM Crops Invade Nature?

Domesticated food crops have had weediness tendencies bred out of their genetic make-up. To date there is no evidence that major food crops, GM or conventional, have become invasive, as shown by extensive investigations, as well as experience gained from a cumulative area of over 300 million hectares of GM crops grown since 1996. There exists no evidence that insertion of a gene for herbicide tolerance or disease resistance can alter the inherent lack of invasiveness. Nevertheless, it is advisable that GM crops are assessed on a case-by-case basis prior to approval for trial or commercial release.

Will GM Crops Outcross with Wild Species?

A range of natural genetic and physiological barriers exists to prevent this. For successful out-crossing, the donor GM plant has to have fertile pollen that can spread, the recipient wild species has to be fertilised and produce viable seed that will germinate to give fertile hybrid plants that can maintain reproduction. Without these conditions being met, there cannot be out-crossing risks to wild species. No evidence has been produced to date to show that GM plants are different in this respect from conventional. GM maize, soyabean and cotton grown in South Africa have no related wild species here with which they can cross.

Transfer of Genes

The technical definition refers to the transfer of genes from a donor species to recipient species with which it is not sexually compatible, with specific reference to transfer between plants and micro-organisms. Such gene flow occurs naturally between bacteria, and the transfer via Agrobacterium to plants is well documented. DNA is digested and destroyed very rapidly. For gene transfer from plants to bacteria, the intact gene has to be in the form of free DNA, be made available to the bacterium, taken up and stably integrated by the recipient bacterium so that it can be functional to express the unique trait. This can only happen when the DNA of the gene and the bacterium have sequence homology.

Extensive investigations have failed to detect such a mechanism of gene transfer in crop plants. A repeated concern deals with the risk of the antibiotic resistance marker genes being transferred to gut or soil bacteria. Detailed assessments have been carried out with the nptII kanamycin resistance gene. Natural resistance to kanamycin has been widespread in both soil and intestinal bacteria long before arrival of GM crops. However, as a precautionary measure, biotech companies are moving away from using antibiotic resistance genes as markers.

Secondary Ecological Impact

As regards non-target soil micro-organisms, one can only refer to extensive studies that could not detect any changes in the microbial population in the rhizosphere region around plant roots between GM and conventional plants. Conventional plant breeding for resistance showed that pests are likely to develop resistance under conditions of high selection pressure. The protein from Bt, turned into a toxin by Lepidoptera larvae, actually has a 40-year track record of persistent efficacy as a spray-on natural bio-insecticide.

The genetic solution to minimising this risk is multifold: using different Bt genes, having a conventional crop as refuge, alternating crops and varieties, varieties with more than one Bt gene, and breeding multiline resistance using GM and conventional systems. Long term use of herbicides on conventional crops has resulted in some weeds acquiring degrees of tolerance. The same could be expected with over-application of herbicides to which GM crops are tolerant and needs to be managed, e.g. using different herbicide tolerant systems.

Impact on Biodiversity

Biodiversity describes the variability within species, between species and in ecosystems. Commercial agriculture has narrowed the genetic base of food crops in some countries. South Africa has had the benefit of having developed most of its own commercial varieties from local and international germplasm, as well as having had access to imported varieties from all continents. Farmers now have a choice of over 3000 plant varieties, more than ever before. Therefore, food crop biodiversity has continued to increase and GM varieties offer additional choices. Farmers are offered a choice. No one is forced to buy GM seed.

The co-existence of various types of maize has been with us for over 100 years: white and yellow maize, popcorn (a snack food), sweetcorn (a vegetable), and special starch types for industrial use. Many countries produce both edible and industrial starch potatoes, as well as industrial and canola oilseed rape. All of these have been managed by way of separate production, handling and marketing. The advent of GM brings another dimension to this equation. The EU has accepted co-existence of GM, organic and non-GM production systems. South Africa is completing identity preservation systems for GM and non-GM crops. There are two sides to this concern: presence of GM in non-GM or organic products, and presence of non-GM or organic in GM products. What is good for the sauce is good for the gander.

Improved seed and advanced plant varieties, in combination with improved farm management systems, fertilisers and crop protection chemicals, have increased South African crop yields per hectare five-fold since 1950 for both maize and wheat. Without this, we would have required 14 million hectares to produce this year's maize crop and 4 million for wheat; a mission impossible. The multi-billion Rand fresh produce export industry would have collapsed long ago without modern varieties, management and quality control systems. The adverse effects of not adopting GM crops need to be assessed. Here are the proven benefits:

— Increased productivity that will reduce pressure on converting forests, wetlands and marginal areas to farming
— Improved net income, especially for small-scale farmers
— Significant reduction of some 60-80% in insecticide use
— Replacing herbicides having persistence in soils and soil water with more environmentally friendly ones

— Facilitating minimum tillage practices that promote mulch and reduce erosion
— Facilitating crop rotation
— Reducing farm worker exposure to insecticides
— Reducing risk of pesticide residues in food products in the face of very strict export market regulations on residues
— Using crops like Bt maize to reduce risk of mycotoxins with known carcinogenic and toxic effects on humans and animals, and helping to comply with strict international standards on mycotoxin presence.
— Using herbicide tolerant crops to reduce risk of weed damage to crops and weed spread into nature.
— Canola and soybean with improved oil qualities to benefit human health
— Perishable fresh produce with longer shelf life.
— Precluding spread of potentially invasive plants can be achieved by conventional and GM technologies to render the pollen or the seeds from crosspollination sterile, as had been done over 20 years ago with sterile lantana.
— Using GMOs for bio-remediation, as well as to reduce industrial contamination and waste.

Organic systems endeavour to eliminate synthetic pesticides and fertilisers, while improving organic content and soil status. An extensive body of scientific data shows that production efficiency is 20-30% less that of the conventional. Organic production could be supported if it benefits the environment over the long term and if farmers obtain sufficient premium to make it economically feasible. What is needed are environmental impact assessments of the effect of ongoing use of neem extract, tobacco extract,

Bt spores or toxin, pyrethrum, rotenone (carcinogenic to rats), and mineral oils as insecticides; lecithin, sulphur and copper salts as fungicides; and farmyard manure, urine slurry, and processed animal products as fertilisers. There is no voluntary or regulatory requirement for ecosystem impact studies prior to use, as for GM crops.

Agricultural Biotechnology

Transgenic crops, main products of agricultural biotechnology, are increasingly becoming a dominant feature of the agricultural landscapes of the USA and other countries such as China, Argentina, Mexico and Canada. Transnational corporations (TNCs) such as Monsanto, DuPont, and Novartis, the main proponents of biotechnology, argue that carefully planned introduction of these crops should reduce or even eliminate the enormous crop losses due to weeds, insect pests, and pathogens. In fact, they argue that the use of such crops will have added beneficial effects on the environment by significantly reducing the use of agrochemicals.

However, ecological theory predicts that as long as transgenic crops follow closely the pesticide paradigm prevalent in modern agriculture, such biotechnological products will do nothing but reinforce the pesticide treadmill in agroecosystems, thus legitimising the concerns that many environmentalists and some scientists have expressed regarding the possible environmental risks of genetically engineered organisms. In fact, there are several widely accepted environmental drawbacks associated with the rapid deployment and widespread commercialisation of such crops in large monocultures

Toxic buildup in GMCs harms useful insects.

— the spread of transgenes to related weeds or conspecifics via crop-weed hybridization

— reduction of the fitness of non-target organisms through the acquisition of transgenic traits via hybridization

— the rapid evolution of resistance of insect pests such as Lepidoptera to Bt

— accumulation of the insecticidal Bt toxin, which remains active in the soil after the crop is ploughed under and binds tightly to clays and humic acids;

— disruption of natural control of insect pests through intertrophic-level effects of the Bt toxin on predators

— unanticipated effects on non-target herbivorous insects (i.e., monarch butterflies) through deposition of transgenic pollen on foliage of surrounding wild vegetation

— vector-mediated horizontal gene transfer and recombination to create new pathogenic organisms

The spread of transgenic crops threatens crop diversity by promoting monocultures which leads to environmental simplification and genetic erosion. History has repeatedly shown that uniformity characterising agricultural areas sown to a smaller number of varieties is a source of increased risk for farmers, as the genetically homogeneous fields may be more vulnerable to disease and pest attack. Farmers have many choices, other than biotechnology, that work. Several people think that HRCs and Bt crops have been a poor choice of traits to feature the technology, given predicted environmental problems and the issue of resistance evolution.

In fact, there is enough evidence to suggest that both these types of crops are not really needed to address the problems they were designed to solve. On the contrary, they tend to reduce the pest management options available to farmers. There are many alternative approaches, (e.g., rotations, polycultures, cover crops, biological control, etc.)

that farmers can use to effectively regulate the insect and weed populations that are being targeted by the biotechnology industry. To the extent that transgenic crops further entrench the current monocultural system, they impede farmers from using a plethora of alternative methods.

World-wide in 2000, transgenic herbicide resistant crops were planted on 74% of the 44.2 million hectares devoted to transgenic crops. 10 In North America, transgenic glufosinate resistant cultivars of canola and corn, and transgenic glyphosate resistant cultivars of soybean, corn, cotton, and canola are now commercially available. Bromoxynil resistant transgenic cotton has also been developed. The so-called Round-up ready soybeans are the most prevalent GMCs.

Transgenic herbicide resistance in crop plants simplifies chemically based weed management because it typically involves compounds that are active on a very broad spectrum of weed species. Post-emergence application timing for these materials fits well with reduced or zero-tillage production methods, which can conserve soil and reduce fuel and tillage costs.

Herbicide Resistant Crops

According to proponents of HRCs, this technology represents an innovation that enables farmers to simplify their weed management requirements, by reducing herbicide use to post-emergence situations using a single, broad-spectrum herbicide that breaks down relatively rapidly in the soil. Herbicide candidates with such characteristics include Glyphosate, Bromoxynil, Sulfonylurea, Imidazolinones among others. However, in actuality the use of herbicide-resistant crops is likely to increase herbicide use as well as production costs. It is also likely to cause serious environmental problems.

It is well documented that when a single herbicide is used repeatedly on a crop, the chances of herbicide resistance developing in weed populations greatly increases. The sulfonylureas and the imidazolinones are particularly prone to the rapid evolution of resistant weeds and up to now fourteen weed species have become resistant to sulfonylurea herbicides. Cocklebur an aggressive weed of soybean and corn in the southeastern US has exhibited resistance to imidazolinone herbicides. The problem is that given industry pressures to increase herbicide sales, acreage treated with these broad-spectrum herbicides will expand, exacerbating the resistance problem.

For example, it has been projected that the acreage treated with glyphosate will increase to nearly 150 million acres. Although glyphosate is considered less prone to weed resistance, the increased use of the herbicide will result in weed resistance, even if more slowly, as it has been already documented with populations of annual ryegrass, quackgrass, birdsfoot trefoil and Cirsium arvense.

Ecological Impacts

Companies affirm that bromoxynil and glyphosate, when properly applied degrade rapidly in the soil, do not accumulate in groundwater, have no effects on non-target organisms and leaves no residues in food. There is, however, evidence that bromoxynil causes birth defects in laboratory animals, is toxic to fish and may cause cancer in humans. Because bromoxynil is absorbed dermally, and because it causes birth defects in rodents, it is likely to pose hazards to farmers and farm workers.

Similarly glyphosate has been reported to be toxic to some non-target species in the soil -both to beneficial predators such as spiders, mites, carabid and coccinellid beetles and to detritivores such as earthworms, as well as to aquatic organisms, including fish. As this herbicide is known to accumulate in fruits and tubers suffering little metabolic degradation in plants, questions about food safety also arise.

Although there is some concern that transgenic crops themselves might become weeds, a major ecological risk is that large scale releases of transgenic crops may promote transfer of transgenes from crops to other plants, which may then become weeds. The biological process of concern here is introgression, that is, hybridization among distinct plant species. Evidence indicates that such genetic exchanges among wild, weed and crop plants already occur. The incidence of shattercane (Sorghum bicolor), a weedy relative of sorghum and the gene flows between maize and teosinte demonstrates the potential for crop relatives to become serious weeds. This is worrisome given that a number of US crops are grown in close proximity to sexually compatible wild relatives.

There are also crops that are grown near wild/weedy plants that are not close relatives but may have some degree of cross compatibility such as the crosses of Raphanus raphanistrum R. X Sativus (radish) and Johnson grass X Sorghum corn.

Agroecosystem Complexity

Total weed removal via the use of broad-spectrum herbicides may lead to undesirable ecological impacts, given that an acceptable level of weed diversity in and around crop fields has been documented to play important ecological roles such as enhancement of biological insect pest control, better soil cover reducing erosion, etc. HRCs will most probably enhance continuous cropping by inhibiting the use of rotations and polycultures susceptible to the herbicides used with HRCs.

Such impoverished, low plant diversity agroecosystems provide optimal conditions for unhampered growth of weeds, insects and diseases because many ecological niches are not filled by other organisms. Moreover, HRCs, through increased herbicide effectiveness, could further reduce plant diversity, favoring shifts in weed community composition and abundance, favoring competitive species that adapt to these broad-spectrum, post emergence treatments.

Environmental Risks

According to the industry, the promise of transgenic crops inserted with Bt genes is the replacement of synthetic insecticides now used to control insect pests. Since most crops have a diversity of insect pests, insecticides will still have to be applied to control pests other than Lepidoptera not susceptible to the endotoxin expressed by the crop. On the other hand, several Lepidoptera species have been reported to develop resistance to Bt toxin in both field and laboratory tests, suggesting that major resistance problems are likely to develop in Bt crops which through the continuous expression of the toxin create a strong selection pressure.

Given that a diversity of different Bt-toxin genes have been isolated, biotechnologists argue that if resistance develops alternative forms of Bt toxin can be used. However, because insects are likely to develop multiple resistance or cross-resistance, such strategy is also doomed to fail. Others, borrowing from past experience with pesticides, have proposed resistance management plans with transgenic crops, such as the use of seed mixtures and refuges. In addition to requiring the difficult goal of regional coordination between farmers, refuges have met with poor success for chemical pesticides, due to the fact that insect populations are not constrained within closed systems, and incoming insects are exposed to lower doses of the toxin as the pesticide degrades.

Non-Target Organisms

By keeping pest populations at extremely low levels, Bt crops can starve natural enemies as these beneficial insects need a small amount of prey to survive in the agroecosystem. Parasites would be most affected because they are more dependent on live hosts for development and survival, whereas some predators could theoretically thrive on dead or dying prey. Natural enemies could also be affected directly through inter-trophic level interactions. Evidence from studies conducted in Scotland suggest that aphids were capable of sequestering the toxin from Bt crops and transferring it to its coccinellid predators, in turn affecting reproduction and longevity of the beneficial beetles.

Sequestration of plant allelochemicals by herbivores which then affect parasitoid performance is not uncommon. The potential of Bt toxins moving through food chains poses serious implications for natural biocontrol in agroecosystems. Bt toxins can be incorporated into the soil through leaf materials, where they may persist for 2-3 months, resisting degradation by binding to soy clay particles while maintaining toxin activity. Such Bt toxins that end up in the soil and water from transgenic leaf litter may have negative impacts on soil and aquatic invertebrates and nutrient cycling processes, all aspects that deserve serious further inquiry.

Downstream Effects

A major environmental consequence resulting from the massive use of Bt toxin in cotton or other crop occupying a larger area of the agricultural landscape, is that neighboring farmers who grow crops other than cotton, but that share similar pest complexes, may end up with resistant insect .

The present serious damage to biodiversity comes from conventional invasive species like woody species, lantana, bugweed and pompom. However, primary threats to biodiversity and ecosystems are population pressure, sprawling urbanisation, industrialisation, uncontrolled commercial harvesting of indigenous plants (an annual R3 billion industry), and inadequate law enforcement.

Recognition should be given to the beneficial role that breeding of improved varieties and GM technology has and is playing in food production efficiency and security, with concomitant benefits to society and the environment. Credit is due to government for efforts to address invasive plants and water quality and especially to the body of dedicated scientists involved in relevant research and application. Efficient enforcement at local grassroots level of increasingly complex environmental legislation remains a serious cause of concern.

Assessing environmental impact of new GM varieties prior to approval for release is to be supported, subject to decisions being science-based. The same concerns are applicable to certain new conventional and ornamental species and alternate farming systems. The endless polarised debate on GM can be short-circuited by having comparative GM, conventional and organic side-by-side impact studies on environment done by a group of credible independent scientists. In the meantime efforts to manipulate concerns about hypothetical GM risks into impending disasters should be exposed for what they are: scare-mongering.

HRCs also have significant problems.

— Reliance on HRCs perpetuates the weed resistance problems and species shifts that are common to conventional herbicide based approaches.

— Herbicide resistance becomes more of a problem as the number of herbicide modes of action to which weeds are exposed becomes fewer and fewer, a trend that HRCs may exacerbate due to market forces.

— Given industry pressures to increase herbicide sales, acreage treated with broad-spectrum herbicides will expand, exacerbating the resistance problem. For example, it has been projected that the acreage treated with glyphosate will increase to nearly 150 million acres. Although glyphosate is considered less prone to weed resistance, the increased use of the herbicide will result in weed resistance, even if more slowly, as it has been already documented with Australian populations of annual ryegrass, quackgrass, birdsfoot trefoil and Cirsium arvense.

— Perhaps the greatest problem of using HRCs to solve weed problems is that they steer efforts away from crop diversification and help to maintain cropping systems dominated by one or two annual species. Crop diversification can

a) reduce the need for herbicides

b) improve soil and water quality

c) minimize requirements for synthetic nitrogen fertiliser

d) regulate insect pest and pathogen populations

e) increase crop yields and reduce yield variance.

Thus, to the extent that transgenic HRCs inhibit the adoption of diversified cropping systems that include rotational crops, cover crops and green manure, they hinder the development of sustainable agriculture. The number of crops engineered for insect resistance is on the rise.

Ecological Risks of Bt Crops

Based on the fact that more than 500 species of pests have already evolved resistance to conventional insecticides, pests can also evolve resistance to Bt toxins present in transgenic crops. No one questions if Bt resistance will develop, the question is now how fast it will develop. Susceptibility to Bt toxins can therefore be viewed as a natural resource that could be quickly depleted by inappropriate use of Bt crops. 15 However, cautiously restricted use of these crops should substantially delay the evolution of resistance. But is cautious use of Bt crops possible given commercial pressures that have resulted in a rapid rollout of Bt crops reaching 8.2 million hectares worldwide in 2000?

Organic farmers can produce a significant yield without insecticides. The refuge strategy of setting aside 20-30% of land to non-Bt crops to delay resistance is very difficult to implement regionally. Data from the Midwest shows that Bt corn saves on some insecticide use and yields are 2.4 Bu/acre higher than conventional corn but only under high European corn borer infestations. On the other hand organic corn growers use no insecticides and obtain yields (4.8-9 t/ha) similar or slightly higher than conventional farmers.

Bacillus thuringiensis proteins are becoming ubiquitous, highly bioactive substances in agroecosystems. Most non-target herbivores colonising Bt crops in the field ingest plant tissue containing Bt protein which they can pass on to their natural enemies in a more or less processed form. Polyphagous natural enemies (polyphagous: subsisting on many kinds of foods) that move between crop cultures are found to frequently encounter Bt-containing non-target herbivorous prey in more than one crop. This is a major ecological concern given previous studies that documented that Cry1 Ab adversely affected the predaceous lacewing Chrysoperla carnea reared on Bt corn-fed prey larvae.

These findings are problematic for small farmers in developing countries who rely on insect pest control, which involves a complexity of predators in their mixed cropping systems. Research shows that natural enemies can be affected directly through inter-trophic level effects of the toxin present in Bt crops. This raises serious concerns about the potential disruption of natural pest control, as polyphagous predators will encounter Bt-containing, non-target prey that move within and between crop cultivars throughout the crop season.

Disrupted biocontrol mechanisms will likely result in increased crop losses due to pests or to the increased use of pesticides by farmers with consequent health and environmental hazards. Toxins from GMCs remain active in the soil, decreasing soil fertility.

Influence on the Soil

The possibilities for soil biota to be exposed to transgenic products are very high. Given the persistence and the possible presence of exudates, there is potential for prolonged exposure of the microbial and invertebrate community to such toxins, and therefore studies should evaluate the effects of transgenic plants on both microbial and invertebrate communities and the ecological processes they mediate.

If transgenic crops substantially alter soil biota and affect processes such as soil organic matter decomposition and mineralisation, this would be of serious concern to organic farmers and most poor farmers in the developing world. These farmers cannot purchase or don't want to use expensive chemical fertilisers. They rely instead on local residues, organic matter and especially soil organisms for soil fertility (e.g., key invertebrate, fungal or bacterial species) which can be affected by the soil bound toxin. Soil fertility could be dramatically reduced if crop leachates inhibit the activity of the soil biota and slow down natural rates of decomposition and nutrient release.

2

Climate-Agriculture Interactions

Weather describes the condition of the atmosphere at a well defined location and at a given time. Climate, on the other hand, refers to average conditions at the same location. Obviously, adjacent areas tend to undergo similar climate conditions and, therefore, climate can be seen as the average atmospheric conditions over sometimes large areas. It is stressed that variability is as much a characteristic of climate as the averages.

Climatic factors like solar energy (radiation) and water are essential to agricultural production; they constitute major environmental resources and the development of technologies for their improved management is one of the roles of agricultural research. A major change in perception that occurred over the recent years is the fact that interactions between atmosphere and agriculture are no longer seen as unidirectional (climate impact), but rather that agriculture also affects global climate as one of the sources of greenhouse gases and a sink for CO_2.

Agriculture is part of the global environment, and that the issue of sustainable agriculture, due to its links with population and socio-economic considerations is thus a relevant topic also beyond the agricultural community, as was highlighted in the recent Kyoto protocol. Climate change impacts can be understood only in the light of current atmosphere-plant-soil interactions, unless unusual or unexpected combinations of factors should develop, where "unusual" refers to changing averages, the frequency of occurrence of factors, their combination and, possibly their nature as in the case of CO_2 and UV-B, two factors which had until recently been regarded as "constants".

A better understanding of current climateagriculture interactions remains thus the key to a better understanding of future conditions and impacts. Large uncertainties are associated with the dynamics of agriculture, as driven by such interrelated factors as population growth, land degradation, international markets, changing diets linked with improving standards of living or other mechanisms, technology and management, etc. Implications for future food security and biodiversity are thus complex.

Climatic Resources

To stress the direct link between agricultural production potential and climate, researchers uses the concept of climate fertility, coined after soil fertility. The fundamental similarities between climate and soil resources include the following: both contribute to the general production potential of a region, both undergo spatial variations and they can be mapped at different scales. In both cases their deficiencies can sometimes be corrected by adequate management practices. In addition, climate and soil contribute to agricultural production potential in an integrated way, not as separate factors, particularly since soil genesis is also very climate dependent.

Climate constitutes a "complex", i.e. set variables which behave coherently, essentially as a result of atmospheric physics and dynamics. For instance, rainfall tends to cool the atmosphere because water evaporation absorbs heat; cloudy days are characterised by a low daily thermal amplitude (difference between day and night temperature), relatively high air moisture and low evaporation, etc.

In addition, the statistical properties of climate derived from long-term observations ensure that the usual range of variation of the "complex" is known. Notwithstanding the difficulties of short-term weather forecasts proper, weather thus behaves rather coherently, and this constitutes an essential piece of knowledge which can be applied to improve the output of agricultural systems in terms of amounts and regularity. The "complex" is also at the basis of agroecological zones (AEZ) interpreted as areas of relative climate stability, associated with typical soils and spectrum of characteristic plants and animals constituting the production system.

FAO has been constantly developing the agroecological zones (AEZ) concept as a basic planning tool in agriculture starting in the late seventies. The "extreme" range, by definition, covers unusually low or high values. The extreme effects on living organisms are far less predictable, for several reasons:

— they may mechanically damage plants and harm animals;

— their occurrence is rare and therefore less data are available for fine-tuning impact models;

— experimental stations tend to discontinue observations after the occurrence of extreme conditions, thereby further reducing the observation base that would be required for impact assessments.

Solar Energy and Light

The sun is the primary source of almost all energy stored

in organisms in the form of biomass chemical bonds, starting with plant photosynthesis. The maximum amount of solar energy available as light depends essentially on astronomic factors, where variations of the solar constant, the annual and diurnal cycles play the most relevant role. Longer cycles (11,000 to 110,000 years) are associated with the orbit of the earth). In general, clouds and various aerosol (sulphates, dust, etc.) can reduce the amount of energy that reaches the ground.

Aerosol, including anthropogenic sulphates, have recently been shown to play a significant part in the energy balance of the earth. Dust is often of volcanic origin and therefore a-periodic, but the increasing frequency of dry haze in West Africa seem to indicate a link with land degradation. Finally, clouds constitute one of the major uncertainties in the current climate change scenarios, as a small increase of the planetary albedo could result in global cooling instead of global warming.

Light also plays an important qualitative role by triggering the photoperiodic response of crops and by influencing the reproductive cycle of animals.

Water

While solar energy sets the maximum value of the energy available for plant growth, water determines to what extent the energy can be used. In fact, plants "pay" for the energy they absorb by transpiring water, and similarly water plays an important part in the thermoregulation of animals. The rather direct and almost linear link between actual water consumption and plant production has been very well documented and the interactions between transpiration and assimilation have been at the core of plant models from the early stages. This simple fact is often overlooked in semi-arid areas: all factors which increase biomass (like the introduction of HYV or the use of fertiliser) also increase water consumption, sometimes resulting in agricultural drought under climatically rather average conditions.

Rainfall still constitutes the main source of water, and many techniques exist to improve water availability through water harvesting, irrigation, flood recession, cropping and grazing, or breeding, for instance for stronger root development. Less conventional sources like dew and fog may be resorted to locally. Finally, air and soil moisture and water bodies play a very direct role in creating conditions favourable for the development of many pests, diseases, pathogens and their vectors (e.g. egg pod deposition and hatching conditions in desert locusts, liver fluke distribution, etc.)

Wind

At the micro-scale, wind has a less direct effect on plants than solar energy and water. It plays a part in mixing air and homogenising temperatures in canopies, thereby ensuring the continued exchange of CO2 and water between plants and the atmosphere. It also contributes to the dissemination of pollen in wind-pollinated plants and the movement of migratory pests, but wind is more often associated with negative effects, including excessive desiccation, physical damage to crops, either directly or through abrasion by transported particles.

Heat and Temperature

Temperature is the yardstick used to measure heat, i.e. the energy stored in the thermal motion of particles. Most chemical processes, and therefore biological processes as well, are temperature dependent. Within the limits of the normal physiological range, the speed of biochemical reaction approximately doubles for a temperature increase of 10°C. It is worth noting that the reactions of assimilation (photosynthesis) are globally less dependent on temperature than, for instance, respiration.

In general, higher temperatures are thus associated with higher productivity and shorter biological cycles. This is amply illustrated by the high biomass turnover of the tropical rainforests (where biomass production is paralleled by high rates of decomposition of plant residues), or the short cycles of pests and diseases in warmer climates.

Many qualitative thresholds are temperature dependent (for instance vernalisation of winter crops, tuberisation, break of the diapause in insects). Finally, air and soil temperatures play a major role in the development of growth of coldblooded animals (earthworms, fish, etc.).

The energy balance and the water balance of crops are interrelated through crop evapotranspiration. Evapotranspiration is a central concept in the determination of the potential biomass, i.e. the maximum quantity of plant biomass which can be accumulated under a given climate, assuming no interference of limiting factors such as poor soils, water stress or pest attacks. Next to quantity, the timing (calendar) of the production is mostly conditioned by limiting water availability in the warm climates, and by temperatures in the temperate ones.

At the global scale, very direct links can be established between the main characteristics of climate and biological productivity. Interestingly, curves very similar to the theoretical biomass curves can be obtained with actual crops, as exemplified below with African crop yields (Figure 1) as a function of National Rainfall Indices. Climate resources also directly affect biodiversity of land and marine ecosystems.

Researchers indicates that the biodiversity relates directly to the number of organisms which can make a living out of available resources, which in turn may be dependent on the extent of these resources, the limitations of the physical environment, the way in which resources present themselves for exploitation (habitat heterogeneity), the number of species geographically available, site accessibility, etc. Forest richness is mainly the product of the resource base—water, warmth and solar energy - : the total available energy is partitioned among species and limits species richness.

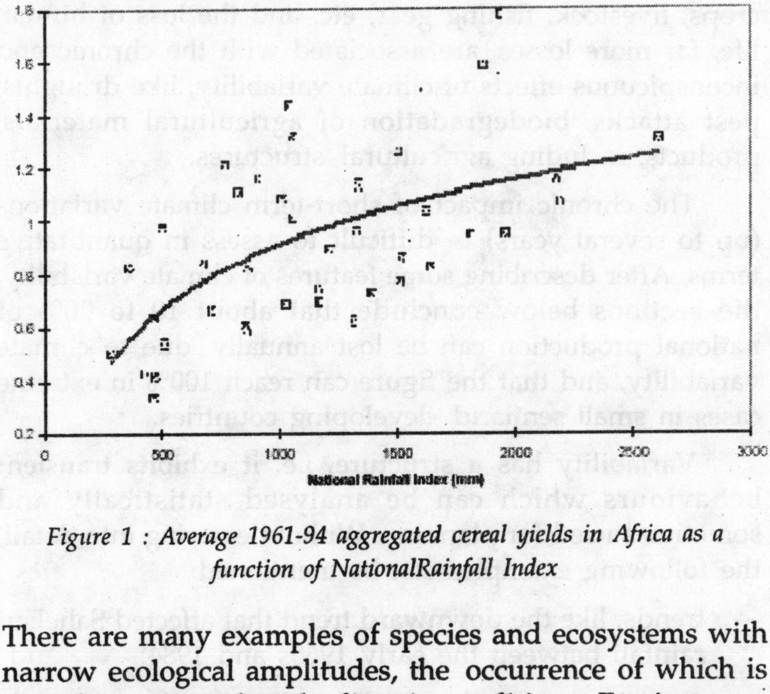

Figure 1 : Average 1961-94 aggregated cereal yields in Africa as a function of NationalRainfall Index

There are many examples of species and ecosystems with narrow ecological amplitudes, the occurrence of which is directly associated with climatic conditions. For instance, on the east coast of the US, the spawning of shad (Alosa sapidissima) peaks at 15°C. Temperature changes could reduce number of repeat spawners and reduce success; mean annual temperature of 11°C appears to be the upper temperature for the formation of ombrotrophic bogs in NW Europe; prolonged summer chilling (less than 4°C) is lethal to the small white butterfly and eggs deposited at temperatures above 33°C are infertile.

Climate Variability

The main hazard associated with weather and climate is random variations, which border on unpredictability at all time scales exceeding a couple of days. The reduction of the uncertainty at the scale of weeks to seasons (seasonal forecasts) would constitute a major improvement of food security. While it is recognised that some extreme climatic factors can provoke massive destruction of infrastructure; crops, livestock, fishing gear, etc. and the loss of human life, far more losses are associated with the chronic and inconspicuous effects of climate variability, like droughts, pest attacks, biodegradation of agricultural materials, products, including agricultural structures.

The chronic impact of short-term climate variations (up to several years) is difficult to assess in quantitative terms. After describing some features of climate variability, the sections below conclude that about 10 to 20% of national production can be lost annually due to climate variability, and that the figure can reach 100% in extreme cases in small semiarid developing countries.

Variability has a structure, i.e. it exhibits transient behaviours which can be analysed statistically and sometimes used for planning. Without entering into detail, the following examples can be mentioned:

— trends, like the downward trend that affected Sahelian rainfall between the early 1960s and 1984;

— persistence, i.e. the tendency for weather types to occur in clusters: runs of dry and wet days, or runs of dry and wet years etc.;

— pseudo-cycles, often a direct result of persistence. A main characteristic of "cycles" seems to be that they collapse before they reach statistical significance;

— extremes follow well known patterns for each of the climate variables.

Several of the listed characteristics have marked effects on agricultural production and food security. For instance, one of the consequences of persistence is that a "bad" year is more likely to be followed by another "bad" year than by a "good" or "average" year. To some extent, when clusters have a relatively predictable behaviour, this information can be used in agricultural planning.

Atmospheric Pollution: Air pollutants affect life in the immediate surrounding of point sources. Some forms of pollution, like acid rain and ozone, are however known to have significant effects on forest and crop yields over wide areas, particularly in humid climates which improve contact between pollutants and plant surfaces. Tropospheric ozone is singled out because it derives for about half from photochemical reactions involving nitrogen oxides and methane, and because of its effect on crops, particularly legumes.

The potential direct negative effects of most climate elements on crops, natural vegetation and forest, and farm animals are well known. They affect all spatial scales from plant organ to the region, through various mechanisms. For instance, high temperatures increase water stress in all organisms; they induce sterility in certain crops, lead to poor vernalisation and increase winter survival of pests in temperate countries. High night-time temperatures are associated with a production loss due to increased respiration loss. In extreme cases, low temperatures can mechanically destroy cell structure (frost); they also contribute to plant desiccation and cause slow growth, particularly during cold waves, etc.

The direct impact of negative climate factors on agriculture, even the spectacular ones (hail, cyclones) is mostly very limited spatially: the main risks lie more with indirect impacts such as conditions favourable for pest and disease outbreaks, fires, etc.

There are many complex interactions where trends affecting non-climate variables eventually lead to a greater vulnerability of production systems to climate variability. Several examples could be given where population growth has lead to the horizontal expansion of agriculture into marginal areas with low water storage capacity: such areas are more prone to agricultural drought as they cannot "buffer out" short dry spells.

Another related factor is the switch from traditional crops grown with low water requirements (millet/ sorghum) towards crops with high moisture requirements (maize) as a result of changing dietary patterns Long-term effects on production and quality of the diet must also be taken into account when trees or plantation crops suffer damage due, for instance, to violent winds or salinisation (due to ocean spray during cyclones, or to other causes).

Climate variability directly and indirectly affects agricultural output (production) through its effects on yields and areas planted. Yields are affected, as indicated, by weather as the main "random" factor, but also by mostly continuous technological trends (including new varieties and management), innovations (including management innovations), agricultural policies (mostly national policies) and extreme factors of various origins. Variation of areas depend more on economic factors.

Areas planted vary according to labour availability, level of mechanisation and expected return (prices). Areas harvested are often strongly linked to environmental conditions, including poor weather during the cycle, damage to infrastructure due to extreme conditions, etc. Because of the complexity of the dependence of area-wide yields on different factors, and because of the high level of aggregation of agricultural statistics, it is not always easy to show the effect of climate variability.

In order to isolate the effect of environmental conditions from other factors, the first step is to eliminate the trends in the series. This makes sense only when there is no marked trend in the weather, which happens to be the majority of cases. In addition, trends are not always constant and the detection of trend changes constitutes, per se, a rather difficult technical problem.

One of the main conclusions of the latest IPCC assessment is that the atmospheric concentrations of greenhouse gases, inter alia carbon dioxide (CO_2), methane (CH_4), and nitrous oxide (N_2O) have grown significantly: by about 30%, 145%, and 15%, respectively. These trends can be attributed largely to human activities, mostly fossil-fuel use, land-use change and agriculture.

Striking difference between the relative importance of CO_2, now the main greenhouse gas (GHG), and water vapour. It is also relevant, to stress that agricultural sources of several GHG are very significant:

— CO_2, an estimated 25% stems from agricultural sources (deforestation: 20%, biomass burning: 5%). The agricultural sources appear to be declining relative to fossil fuel use;

— CH_4: there is a lot of uncertainty about the figures. About 70% could be derived from anthropogenic sources, of which 20% each each domestic ruminants, biomass burning and rice production. The remaining 10% is usually assigned to "other waste products". Natural wetland could be responsible for about the same emissions as rice fields;

— N_2O, tillage 44%, fertiliser 22% and biomass burning 9%. Agricultural sources seem to be stabilising, while the relative importance of energy is on the increase.

The scenarios developed by IPCC present the following data for 2100: + 1°C with 500 ppmv CO2 (lowest), + 2.0

to 2.5 °C with 725 ppmv CO_2 (most likely) and + 3.5 to 5 °C with CO2 above 1000 ppmv (worst case). According to Ehsan Masood, if all developed countries kept to their Kyoto target, world temperature would still rise by 2.1°Cby 2100. This is only 0.27°C lower than the "Business as Usual scenario" (BAU) resulting from no intervention at all to reduce emissions. This is to say that about 2°C increase by 2100 can be taken for granted.

Under the "virtually certain 'facts' to very probable 'projections'", Mahlman lists the following:

— climate has considerable inertia, and cannot be reverted over a short period of time;
— cooling effect of sulphate particles is insufficiently quantified;
— significant reduction of the uncertainties will require a decade or more;
— water vapour concentrations will increase in the lower atmosphere, and global mean precipitation could increase by 1.5 to 2.5 % per 1 °C of global warming;
— sea level rise may reach about 50 cm by 2100.

There is no scientific evidence that the frequency of tropical cyclones, storms or hurricanes would increase, nor should winds in mid latitudes.

Improving Climate

Climates can be improved using a variety of traditional and modern methods and techniques, from microclimate manipulation (such as mulching), the orientation of the rows in row-crops and windbreaks through varying levels of water control to artificial climates, for instance in greenhouses where temperature, light and CO_2 can be controlled.

While the techniques like those of frost protection using smoke (fires) and sprinkler, or hail protection and cloud seeding are very "hardware oriented", there is currently a tendency towards more sophisticated use of weather knowledge in models that assist farmers and herders in decision making. Strictly speaking, this is not (micro)climate manipulation, but a way to make optimal use of climate resources.

The dependence of many common pests and diseases on weather can be modelled and their probability of reaching critical levels can be forecasted with good accuracy. However, many farmers and livestock breeders carry out preventive control of pests and diseases, thereby avoiding almost all losses but also, and maybe more importantly, actually insulating their production from weather, and ensuring a better control over their time and capacity to plan.It remains that, at least statistically, agrometeorological pest and disease models have the potential to reduce protection costs while reducing pesticide release into the environment.

Taking advantage of climate knowledge to manage variability The first step in all climate-agriculture work is to know the climate, i.e. to collect long-term data and assemble them into climatic databases. But very often there is a mis-match between biological observations and weather observations: efforts should be made to collect at least basic information from the location of all agronomic experiments, and to ensure that extreme conditions are adequately measured.

Climate plays a part at the many different levels leading from production or the harvest of wild products (including fish) to consumption, and the effect of weather and climate on all those levels is worth investigating. It should also be kept in mind that the definition of extreme agrometeorological events is broader (than just weather), as they include as well weather conditions conducive to the development of agents (like pests and diseases) that negatively affect agriculture.

Extreme agrometeorological events thus include, for instance, desert locust outbreaks: rainfall in semi-arid areas can create conditions favourable to locusts, eventually leading to gregarisation and swarms spreading over large areas. Fire as well can be included, as risks are very dependent on rainfall, moisture and winds.

The two previous examples stress the need to carry out both a monitoring of conditions favourable to pest, disease, fire, etc. development and, subsequently, to monitor the conditions which control the spread of the extreme agrometeorological event from its source to larger and sometimes distant areas. Schematically, the use of climate knowledge is relevant for the following "categories":

— types of "organisms": crops, pollen and seeds; livestock, poultry and freshwater fish; pests, diseases and parasites; national parks and wildlife;

— monitoring growing and post harvest conditions. This can be done by using models, identifying critical thresholds, forecasting the impact of drought, fires, hail, frost, vernalisation, extreme agrometeorological factors;

— advice, as basis for decision making: meteorological forecasts, trafficability, farm operations, dissemination of pathogens and pollutants, response farming and precision farming, advice to farmers and herders, irrigation and drainage;

— technology (climate as resource): water conservation and harvesting, occult precipitation, artificial climates (greenhouses, stables, tractors and farmhouses), agroclimatic zoning, agroclimatic risk (insurance), microclimate modification, windbreaks;

— forecasting: yield forecasting, forecasting quality of production, forecasting phenology (harvest time and labour requirements), rangeland production, international markets as a result of production and resulting prices;

— methodologies: data technology in the broadest sense (collection, standardisation, software, estimation of missing data, random weather generators, gridding), modelling production, development, impacts.

Forecasting in general deserves a special mention in the light of the interest triggered by the latest 1997/98 El Niño/Southern Oscillation, in particular because of the lessons than can be learned. To start with, El Niño provides a good example of a relatively reliable seasonal forecast, which can be taken advantage of by governments and the agricultural community alike.

Unfortunately, like all forecasts, the actual magnitude, location and other features are affected by error. The methodology of whether to take management decisions, taking into account ENSO information, must be based as much on economic data as on simulated potential agronomic impact scenarios.

Such impact scenarios are not ENSO specific but constitute only a special application of a weather or climate forecast. Current discussions about how to react to El Niño are useful only if they lead to long-term solutions, in particular if efforts are made to improve all seasonal forecasts, if decision/simulation tools to be used by governments and farmers and ranchers are available, and if climate/weather impact on agriculture is seen as much in terms of opportunities and more efficient use of climate resources by farmers rather than only in terms of loss mitigation.

The decision tools must be developed in collaboration by national climatological services, research and agricultural and livestock extension and tested locally, including a critical evaluation of the impact of the advice on agricultural output in terms of quantities and regularity. In practice, the decision tools are tables/flow-charts or software that assist farm-level management decision-making based on three types of inputs:

— the knowledge of local environmental/agricultural conditions (reference data);

— the measurement of local "decision parameters" by local extension officers, farmers and cattle breeders;

— economic considerations, e.g. cost of inputs vs expected output.

Adaption of Species and Production Systems

There are many examples of adaptation of agriculture to extremely difficult climates, for instance, flood recession crops on river banks and irrigation in semi-arid areas and outright deserts. Another technique uses low mud dams (usually not exceeding a height of 1.5 meters) in terrain depressions to collect run-on water during the very short seasons of the desert margins. As water infiltrates and evaporates, crops are planted at the receding edge of the water and grow on soil moisture during the dry season, thereby enjoying low pest incidence due to low atmospheric moisture and a high production potential due to the virtual absence of clouds. In addition, the harvest is spread over several months, and is predictable.

Livestock is another convenient mechanism to collect the sparse biomass spread over large areas. In semi-arid areas it constitutes virtually the only mechanism for people to make a living in spite of very limited water resources. More modern techniques are those of precision farming, though still beyond the reach of most developing countries, which adapt management to the micro-variations of the environment, in particular soil features.

— the breeding of plants and animals better suited to resist abiotic stresses (like high temperatures) through inter-specific crosses

— the development of morphological adaptations - such as stronger root development and more favourable Leaf Area Index—and physiological improvements to photosynthetic efficiency—like possibly CAM features in C_3 and C_4 plants

— more specific pest and disease resistance.

Conjectured effects of climate change on agriculture are large, but uncertainties are such that specific protection measures (e.g. location- or species-specific) are unlikely to achieve their goal. It is rather through emission reductions, gradual adaptation (including evasion) and preparedness that agriculture will be able to cope with the new environmental setting. Protection is likely to involve structural measures with a sizeable cost.

The uncertainty associated with projections of climate change and assessments of impacts on agricultural potential calls for attentive preparedness, to readily take advantage of beneficial impacts of climate change and increased atmospheric CO_2, to mitigate negative impacts of climate change where they cause loss of productivity, and to cope with the technological and social challenges of changing patterns of land productivity.

In essence, this will require addressing many problems which concern farmers, foresters and livestock breeders and decision makers already today. The prevailing philosophy, particularly in developing countries, has been "no-regrets", i.e. only measures that make economic sense now should be adopted, because they reduce emissions from the agricultural sector or improve resilience of all sectors of agriculture against weather variability. All have a marked management component and could thus often be implemented at minimal cost. Such measures include:

— improved fertiliser use, as N_2O released into the atmosphere is a loss and constitutes a symptom of inefficient farming. Needless to say, the same applies to nitrates lost to the water table and surface waters;

— improved ruminant digestion through more efficient feeds or, when feasible, a shift to enzymic digesters;

— development of water harvesting and conservation techniques, as well as other improvements to crop-water management as an adaptation to rainfall variability;

— improved rice farming, as higher yields are accompanied by a relatively smaller loss of methane;

— improved soil carbon storage (carbon sink) while at the same time improving soil structure increasing water holding capacity;

— improved low-impact harvesting in forests, reduction of slash-and-burn agriculture, better soil protection;

— a growing of alternative energy crops.

The example of rice provides an interesting illustration of the fact that agricultural emissions of greenhouse gases should be linked to production rather than expressed in absolute amounts, in order to favour the most efficient production systems. Finally mention energy crops as a way to close the carbon cycle, as well as carbon sequestration in soils and biomass. While modern fuels from biomass seem to have real potential in reducing fossil fuel consumption while generating income in rural areas, carbon storage in biomass often needs justifications which are not all relevant for agriculture, such as recreation and tourism, watershed protection, improvement of urban climates.

Several areas have received insufficient attention in the general field of climate-agriculture research, including both climate variability and change. They include the following:

— the collection of combined data-sets of climate and agronomic information needed for the development of realistic impact assessment tools for current and future climate. The agronomic information should include at least information on varieties, phenology, pests and diseases, management and yields. This is also a point noted by Fischer et al. when they suggest that agricultural research would benefit from greater attention to macro- and micro-climate in all trials;

— efforts should be made to improve climate (i.e. seasonal) forecasts. While this is a task for climatological research, there are already some cases (El Niño) where the forecasts have achieved a reasonable degree of reliability. It is up to agronomic research to determine the conditions under which the use of the forecast will improve sustainability of all types of agriculture (from crop agriculture to inland fisheries) in terms of economic return and food security;

— many global impact assessments are based on models designed for the level of the individual field. They are appropriate for farm and forest management, but inadequate for regional studies, if only because most of the required inputs are meaningless or impossible to obtain at regional scales. Methods are needed that will allow to bridge the gap between the field and regional scales, as well as between "sectors" (e.g. combined rangeland species composition and livestock models). This point is highlighted by Leemans and van den Born who stress that detailed process-oriented models are unsuitable for input data on a coarse spatial grids;

— the potential interactions between changed climate and other plant responses, for instance to mineral nutrition need to be further investigated;

— equilibrium global circulation models provide little insight into the dynamics that will lead from the current situation to future situations 30 or even 100 years from now. Transient models should be used instead and tested against recent historical data. A model that cannot represent the current situation is unlikely to be useful in forecasting future situations;

— the question of rates of change: a thorough discussion of impacts on natural vegetation and landscape processes as function of rates of changes is still missing, as well the links between ecological complexity and resilience.

Climate-Driven Activities in the CGIAR System

Much research work carried out in the CGIAR system bears a direct relation to climate and climate change:

— Breeding for yield stability in stressful (drought, cold, heat) and variable environments: maize and wheat, barley and durum wheat, legumes, maize; breeding for earliness: spring wheat (ICARDA), maize (IITA); breeding for increased productivity in flood-prone lowlands (IRRI). WARDA has a holistic approach to "drought resistance", a mix of physiological, phenological and morphological mechanisms for escape,. avoidance, resistance and recovery. Breeding should, therefore, be environment specific;

— Models to simulate management:

  — improvement of water use efficiency, model research efficiency and outcomes in desert margins systems;

— Improved decision making in several areas:

  — use of water resources: water conservation and management, comparison of agroforestry and crop agriculture in respect to sustainable water use, practices and breeds which improve water-use efficiency of rice;

  — coastal zone management, fisheries and aquaculture

— Climatic and land-resources databases, and conservation of resources: agroecological characterisation; aquatic biodiversity and coral reef conservation; baseline data on above-ground biodiversity

— Climate and seasonal forecast: IFPRIs medium-term plan proposes "exploratory research" on economic and agricultural implications from medium to long-term climate forecasts;

— Management techniques to reduce CH_4 emissions from rice fields, development of drought resistant upland rice;

— Management strategies: research into the effect of changing weather patterns on floodprone riceland; improved-impact harvesting of natural forests; alternative land-use strategies and evaluation of international climate change mitigation policy options.

Furthermore, several centres have established scientific links with relevant organisations such as IGBP, in particular with the Global Change and Terrestrial

Ecosystems (GCTE) project Focus 3 dealing with agriculture, forestry and soils. The main items covered include the effect of global change on key agricultural systems, including food crops and pastures, changes in pests and diseases (with emphasis on weeds), effects on soils, on multi-species agro-ecosystems and on managed forests.

There has been long-standing collaboration between FAO and WMO on climate and agriculture, often jointly with UNEP, UNESCO in the Interagency Secretariat of Agricultural Biometeorology, etc., directly or through the WMO Commission on Agricultural Meteorology. Bilateral contacts of FAO with several CGIAR institutes have mostly centred on agroclimatic data and the co-sponsorship of several meetings on the agrometeorology of specific crops, starting with rice, sorghum and millet, groundnut, or the characterisation of agricultural environments.

The Kyoto Protocol explicitly recognised the links between climate change issues and sustainable development. In view of the new commitments and options, an increased role of the international agricultural community can be envisaged:

— participation in the development of improved assessment techniques for emissions and sinks of greenhouse gases in the agricultural sector, including terminology issues (e.g., managed and natural "forests");

— participation in the discussion of details of trading emissions, its links with the clean development mechanism, and tools for developing countries to benefit from the trading with emissions;

— assistance to countries in their compliance with new obligations deriving from the UNFCCC and the Kyoto protocol.

3

Ecological Assessment of Food Technology

Since the beginning of history, man has used different types of technologies for storing and processing the foods. But improper storage and incorrect and/or excessive thermal processing are known to affect nutrient content in foods. Vitamins are more susceptible than minerals, and amino acids are somewhere in between. On the other hand, appropriate storage conditions and well-controlled processing have positive effects in retaining the original nutrient content of the food, and in many cases improving the availability of specific nutrients and the overall quality of the product.

Food legumes present a variety of problems, from production to consumption, that will be used to point out the effects of storage and processing on nutritive value. Common food legumes, in particular common beans, are harvested in periods of the year when there is rainfall, thus they must be dried for storage. The removal of excess moisture by sun-drying, if not controlled, may initiate a condition known as hardening, which makes the beans hard to cook. This is accentuated during storage at high temperatures and high relative humidity.

Whether hard to cook or not, beans are milled in certain areas of the world and in others they are thermally processed whole for consumption, mainly to inactivate antiphysiological substances and improve nutritional value. Other processes are also used, such as germination and fermentation. If not properly stored, processed products may lose significant portions of their nutritional value.

Post-harvest Processing

Storage

Processing for storage as grain. After harvest, common beans usually contain relatively high amounts of moisture that has to be removed before storage. It is a common practice carried out by small farmers in developing countries to expose the grain to solar radiation. The exposure time is more or less well controlled; otherwise the farmers have indicated that the beans will become hard. As solar exposure time increased, there was an expected decrease in moisture content, but at the same time, water absorption and cooking time decreased.

Changes in other chemical constituents also took place, the significance of which is not well understood; however, this has important implications in the subsequent operations involved in taking beans to the consumer. These changes probably have an influence in the hard-to-cook condition often developing in common beans and other food legumes subjected to storage. The hard-to-cook phenomenon. Storage may have direct and indirect effects on the nutritive value of foods and diets.

Poor storage conditions with high temperature and relative humidity will result in staple foods with high moisture levels and a decrease in quality because of the Maillard reaction. Such conditions will also result in the growth of fungi that produce toxic compounds that have adverse effects on animals consuming such foods. Poor storage conditions can also affect nutritive value by favouring insect infestation, which also results in losses of dry matter. Finally, the improper use of chemicals to protect grains can also decrease the nutritive value of the staple food. Various examples will be given for situations that are common in tropical developing countries.

Poor storage of common beans will increase the hard-to-cook problem. At the end of each storage period, samples were cooked under standard procedures at atmospheric pressure and their hardness was measured by the Instrom texturometer. A standard cooking hardness at g-force of 90 was chosen on the basis of current home practices in Guatemala.

However, the samples at 3 months required 170 minutes, while those stored for 6 months were still uncooked even after 210 minutes. This hard-to-cook condition has significant economic implications, not only because the beans are no longer acceptable to the consumer, but because if they are cooked, the increase in energy cost is very high. The loss in PER and weight is the result of lower protein availability as well as a loss of Iysine, which becomes inactivated through the well-known Maillard reaction shown by various investigators.

An interaction between storage and processing can also affect the nutritional quality of legume foods. In some instances long cooking time is needed to obtain a maximum possible nutritive value from hard-to-cook beans. High moisture in the grain favours hardening as storage time increases. The mechanism of the hard-to-cook condition that develops in food legumes under improper storage conditions is not fully understood.

The evidence available suggests that an increase in the bound protein takes place in the seed coat and aleurone layer; however, the cotyledons also lose their capacity to absorb water because of changes in pectins and calcium ions, and very often they develop a grey colour, suggestive of carbohydrate-protein reactions.

Insect infestation. Some studies indicate that chickpeas or pigeon peas lost about 18 per cent of their protein quality because of insect infestation. The loss may have been caused by contamination from uric acid, a protein metabolite of insects, as well as by increases in fat acidity and microbial contamination, and perhaps even to losses in grain fractions. Resistance to insect attack has been reported for common and Faba beans. It is possible, therefore, that natural resistance to insect infestation exists, and thus chaser cooperation should be established to identify the nature of this resistance in order to increase it to alleviate storage problems, reduce chemical treatment for insect control, maintain nutritive value, and insure increased efficiency of processing.

Processing

Milling

Most staple foods as well as other foods of vegetable and animal origin are processed for consumption. Although not universally used, in certain regions of the world, too legumes are milled to remove the seed coat before the application of thermal processing. Storage conditions have significant quantitative and qualitative effects on legume milling. Milling of food legumes is carried out by household and commercial methods involving the same basic principles. There is both a wet and a dry method. Both procedures have disadvantages. The wet method usually takes longer, and although yields are good, the cotyledons (dhal) become hard to cook and require longer cooking.

The dry method has the disadvantage of high milling losses caused by breakage of cotyledons and Powdering.

Furthermore, the loosening of the seed coat in this process is not adequate, and even less so when the drying operation of beans before storage has not been satisfactory. In addition, if the grain is insect-infested, milling yields decrease. "Control" chick peas with 2 per cent kernel damage had an 82 per cent dhal yield, while infested peas had 15 per cent kernel damage and a 65 per cent dahl yield. Milling techniques have been developed to maximize the yield of edible fractions, and data have been obtained suggesting that the genetic make-up of the cultivars also plays an important role in yield of edible fractions.

Cooking

Roasting: Roasting is an interesting processing technique because it has the capacity to develop attractive flavours in foods so treated. It also induces important functional properties, attributes that should be compatible with nutritional value. Length of roasting is important in enhancing PER in these bean's. Chemical analysis for available Iysine showed the expected decrease in this amino acid, which explains the lower protein quality as roasting time increased. This product, converted into a fine flour, is used as a drink for young children in some developing countries

Based on the components measured, roasting resulted in a product as good or better in protein quality as that produced by common wet cooking under pressure. Therefore, if processing conditions are well controlled, there is no danger of reducing the nutritional potential of a particular food. Heat treatment can also be very useful as a way to pre serve the cooking characteristics of whole food legumes that become hard to cook upon storage, as indicated earlier.

Cooking at atmospheric pressure and under pressure: Wet cooking is probably the most common method of preparing food for consumption, both at home and industrially. Common beans must be cooked before consumption in order to inactivate antiphysiological factors. The antiphysiological factors are destroyed in about 90 minutes, which results in an increase in protein quality. However, excessive cooking time causes a progressive decrease in nutritive value because of the loss of Iysine. Wet cooking can be carried out under pressure, which reduces the time of exposure compared to cooking at atmospheric pressure. In both situations, however, excess cooking time may result in lower nutritive value.

Drum drying: Another technique for cooking foods is drum drying. Important processing conditions to inactivate antiphysiological factors and up-grade the nutritional quality of the product include temperature, residence time, both dependent on drum velocity, opening between drums, and solid concentration. It is an attractive technology because the material is cooked and dried in one operation.

Germination

It is generally known that germination markedly improves the nutritional quality of legumes. A dramatic increase in ascorbic acid in legume seeds has been observed during germination. Also, the concentration of a number of other vitamins is high in germinated legume seeds. Various attempts have been made to study the effect of germination on the reduction of antiphysiological factors and changes in the organic constituents of legumes in order to improve their nutritive value, but contradictory results are reported in the literature, particularly with respect to protein quality.

Most of the reports on the subject indicate that trypsin inhibitor activity is retained in germinated food legumes without affecting nutritive value, an observation deserving more research. Likewise, starch is broken down and both flatulence factors and polvphenolic content are reduced. Germination did not affect protein quality significantly, although there was a small increase in digestibility. This should have led to an increase in protein quality, but it did not occur because bean protein is deficient in sulphur amino acids. interestingly, trypsin inhibitor activity did not decrease with germination time; however, this did not affect protein quality.

Fermentation

Processing food by fermentation has been practiced by man for centuries and has been used quite extensively in various parts of the world, particularly in the Orient. The number of foods that normally undergo fermentation is relatively high, and there are obviously many changes in chemical composition and nutritive value, all well documented. Some general observations made by different workers on the effects of fermentation on nutritive value include increases in B12 as well as in other vitamins of the B group. Likewise, there are increases in protein quality, increased availability of various nutrients, and removal of antiphysiological factors.

A mixture of soybean and wheat in a ratio of 60:40 provides protein of a quality higher than that of either one alone. If these foods individually or combined are fermented through the use of micro-organisms such as Rhizopus oligosporus, there are important increases in protein quality. The fermented product in every case showed a higher protein quality, depending on the inherent protein quality of the food before fermentation.

For example, wheat protein quality increased much more than that in soy because the protein of the micro-organisms contains lysine, which is deficient in wheat, and both micro-organisms and soy protein are deficient in methionine. The point is that fermentation offers very good opportunities to develop foods of higher quality. Finally, fermentation processes using solid substrates have been considered as an attractive alternative for the production of new sources of protein foods in the underdeveloped countries.

4

Green Revolution

Green Revolution is the term used to describe the spread of new agricultural technologies that dramatically increased food production in the developing world beginning in the middle of the 20th century, the impact of which is still felt today. In the 1940s, to address the problem of impending famine from a growing imbalance between population and food supply, the Rockefeller Foundation provided funds for agricultural advances in developing nations and gathered together a team of dedicated researchers from various parts of the world. Other national and international institutions joined in this effort over the following decades.

The crops that were developed as part of this effort were superior to other locally planted crops in yield increase, yield stability, wide-scale adaptability, short growing season duration, resistance to biotic stresses (diseases and insects), tolerance to abiotic stresses (drought and flooding), and grain quality. Importantly, fertilisers, pesticides, and irrigation systems were also introduced in many parts of the developing world, and this whole technology package was supported in many parts of Asia and Latin America by subsidies and government-guaranteed product prices. The Green Revolution has been credited with "making an astounding jump in basic food crops in the developing world, especially Asia". It significantly increased food production in Asia and Latin America at a time when massive malnutrition in those areas was feared. Its success was due to a combination of circumstances that made the time ripe for an agricultural revolution to take place in much of the developing world.

Worldwide, policymakers saw increased food production as a priority after World War II and set the wheels in motion to facilitate that goal. At the same time, agricultural science and technology in the industrial world-which produced hybrid seeds, pesticides, fertilisers, and irrigation systems-had advanced to the point that new technologies could be transferred to other parts of the world where they could be useful. Indeed, universities in the United States played a crucial role in the advancement of agricultural development in that they were a source of scientists and a provider of education, particularly advanced degrees.

Two of the criteria for an agricultural revolution are a substantial improvement in food production and sustainable use of the relevant technologies. The Green Revolution, in the many places that it was adopted, met these criteria very well. What made the Green Revolution technologies successful in bringing about revolutionary change was the integration of three factors: (1) plant breeding methodologies that were adapted in innovative ways to produce regionally useful crops, (2) scientists and researchers who combined a range of various technologies to achieve their objectives, and (3) local scientists who were given the necessary training to breed the crops themselves. These three factors, along with programmes and policies that kept farmers informed and protected their production incentives, were responsible for a significant increase in food production.

Plant Breeding

The plant breeding research that led to high-yield varieties (HYVs) of staple food crops was perhaps the most important factor in the food production increase. Several genetic traits were selected to increase yield, yield stability, and wide-scale adaptability of rice, wheat, and maize. Three types of breeding strategies were used to develop these three types of crops.

The first was conventional breeding, in which parents of the same line were crossed and the resulting offspring were screened for desirable traits, and those offspring that contained the traits for the next round of crossing were selected. The second strategy was hybrid breeding, in which plants from different lines were crossbred to produce offspring with increased vigor. The third strategy was the innovative technique of shuttle breeding, which at the time violated the conventionally held belief that plants had to be adapted to the native soils and climate in order to be successful, and only if all selections were from plants that grow in the actual area where the crossbred crops are to be produced could the new varieties be well-adapted to their environment.

Shuttle breeding, in contrast, involves growing plants at various locations and crossbreeding between those lines to obtain offspring with widely adaptable traits that can be grown under diverse growing conditions, and, more specifically, with improved resistance to the devastating disease of wheat rust. As a result, new high-yielding varieties of wheat, maize, and rice were developed around the world with a better ratio of grain to straw, shorter and sturdier stems, and better response to fertilisers.

The combination of irrigation systems, fertilisers, and pesticides, along with the HYV seeds, was also key to the increase in food production. When the HYVs were grown with these other inputs, they produced substantially higher yields. Aside from increased yield in most situations, another benefit that the combined Green Revolution technologies provided was greater yield consistency in the face of differing agricultural conditions. For example, the Centro Internacional de Mejoramiento de Maizy Trigo (CIMMYT) developed a maize variety that gave the second-best yield in Mexico when external planting conditions were optimal but inputs were less than optimal.

However, even in times of drought, the CIMMYT variety, combined with the appropriate chemical inputs, provided by far the best yield. Table 1 illustrates the changes in adoption rates of various Green Revolution technologies-acreage and percentage of new highyield crop varieties, irrigation, fertiliser consumption, number of tractors, and cereal production-in developing countries in Asia.

Table 1. Changes in Production Inputs in India, Pakistan, and China Due to the Green Revolution

	Adoption of High-Yield Wheat Varieties (million ha/% of total growing area)	Adoption of High-Yield Rice Varieties (million ha/% of total growing area)	Irrigation (millions of ha)	Fertiliser/ Nutrient Consumption (million of tonnes)	Tractors (millions)	Cereal Production (million of tonnes)
1960	0/0	0/0	87	2	0.2	309
1970	14/20	15/20	106	10	0.5	463
1980	39/49	55/43	129	29	2.0	618
1990	60/70	85/65	158	54	3.4	858
2000	70/84	100/74	175	70	4.8	962

Local scientists were given the necessary training to breed the crops themselves so that they would be able to advance the agricultural revolution in their own nations without continued supervision from the outside. The Food and Agricultural Organisation of the United Nations (FAO) and the Rockefeller Foundation established a training programme that recruited young scientists from around the world to learn the relevant agricultural disciplines. After their training was completed, the scientists were given HYV semi-dwarf seeds to take back to their respective nations.

As a result of these successful new technologies and their transfer, food production improved enormously over the course of only a few decades in many parts of the world. Taking Latin America as an example, based on the combination of HYV wheat varieties, chemical inputs, and expanded acreage, total maize production increased by 142 percent and total wheat production increased by 100 percent between 1961 and 1991. Even earlier, Mexico was able to achieve self-sufficiency in wheat and maize production by 1956 through Green Revolution technologies.

In Asia, HYV wheat and rice were first tested in the early 1960s, and production campaigns were launched soon afterward. These campaigns led to a sharp increase in adoption of HYV wheat and rice, yields, and production. In South Asia, for example, rice production more than doubled from 1961 to 1991, and wheat production almost quadrupled in that time.

How the Green Revolution was financed can be seen as a twofold process: funding the research to develop the technologies and to transfer them to various parts of the world, and then financing the agricultural technologies in those regions so that the local farmers could afford to use them. International and domestic funders (i.e., funders within nations where Green Revolution technologies were being promoted) were key to this process. International funders acted first to provide the necessary monetary, scientific, and human capital, and domestic funders cooperated with these international groups to sustain the agricultural movement within particular regions.

Importantly, the international sources of finance were public organisations, both governmental and nongovernmental, acting ostensibly out of public interest rather than from business-related motives. Funders and scientists being involved in helping other nations was a rather new concept at the time. Prior to World War II, there was little support for nations working together to provide finances for worldwide industrialisation and agricultural modernisation. An important exception was the Rockefeller Foundation, which already had a programme in China in the 1920s to improve wheat production.

But the outbreak of World War II changed the philanthropic programmes of the Rockefeller Foundation as well as those of other funders; as such, their efforts to improve agriculture took a global turn. Public-sector funding and leadership turned out to be crucial to ensuring the success of the Green Revolution. The Rockefeller Foundation in the 1940s is generally regarded as the "founder" of the Green Revolution, with its programmes committed to agriculture in the developing world. Later, the World Bank; the FAO, and the United States Agency for International Development (USAID) joined in the effort.

At the country level, success was usually due to collaboration between the international agricultural research centers (IARCs), governments, and their national agricultural research systems (NARS). The IARCs gathered test nurseries of selected seeds, distributed them widely to the NARS, and analysed and shared the results of the seed development. NARS, thus, had in their hands a wide selection of germplasm for use in their own breeding programmes from which to develop locally adapted HYVs. Indeed, an important factor that contributed to the sustained success of the Green Revolution was that scientists received financial support from these public institutions to set up research centers in developing nations. As such, they had firsthand information on what the local farmers needed and what the local planting conditions were.

Furthermore, they were on site to train local farmers and scientists to ensure that knowledge of how to use these new technologies would be passed on. Recognising the importance of developing local expertise, organisations in developed countries pooled their resources to aid in the effort to link industrial-world scientists with developing-world locations, which resulted in the formation of the Consultative Group on International Agricultural Research (CGIAR). CGIAR now includes a system of 16 Future Harvest Centers working in more than 100 countries conducting research on all key crops of the developing world and on livestock, fish, forestry, plant genetics, and food policy.

With the creation of CGIAR, support for the Green Revolution became more broad-based and included European nations, Canada, and Japan. Now some 50 donors, including the World Bank, the United Nations, the United States Agency for International Development, and other public organisations, have provided expanded funding for the Green Revolution through and in association with CGIAR. Farmers in Latin America and Asia were able to afford the relatively expensive HYV seeds, pesticides, and fertilisers through a combination of internationally provided subsidies and low-interest loan systems. Adoption of the Green Revolution package was relatively expensive; in Bangladesh, for example, the input costs were 60 percent more than those for traditional seed varieties.

Small subsistence farmers could afford these packages only if they borrowed money, which usually came at a high interest rate. Establishment of rural banks was important for adoption of the Green Revolution technologies in Asia and Central America. The substantial production improvements that farmers experienced through these Green Revolution technologies enabled them to pay off their loans quickly and easily. Moreover, the money earned from the increased food production often meant that whole villages could make investments in their schools and roads, efforts that in the long run would further improve not only agricultural output but also many other aspects of a community's socioeconomic well-being.

The Green Revolution is generally considered to have been a success in four distinct regions of the world: Latin America, China and Southeast Asia, India and South Asia, and the United Kingdom (UK). Except for the UK, success was generally defined by a production increase that staved off potential malnutrition, quite apart from concerns about the environment, socioeconomic equality, or other such issues. Millions of lives were at risk from malnutrition in Asia and Latin America immediately after World War II, and results were needed, fast. An increase in food production had to be the top Green Revolution priority in those areas.

Latin America was a primary region of interest during the Green Revolution because of a combination of need and local governmental interest in improving agriculture. The European conquest and its centuries-long aftermath had resulted in depletion of soil nutrients and other natural resources that were necessary for sustainable crop yields. The Rockefeller Foundation created the Mexican Agricultural Programme (MAP) in 1943 amid concerns about political stability and national security as well as the specter of malnutrition.

The Ford and Rockefeller Foundations and the Mexican government signed a memorandum of understanding in Mexico in early 1943 to establish the Office of Special Studies (OSS), a follow-up to MAP, as a semi-autonomous research unit within the Mexican Department of Agriculture. Until its transformation in 1966 to the CIMMYT, the OSS was the research center that spurred a major transformation of Mexican agriculture, increasing yields in wheat and maize to an extent that effectively prevented widespread famine.

The Green Revolution also achieved significant agricultural yield increases throughout Asia. Before the new agricultural technologies and practices were brought to the continent, wheat yield in India and China was on par with that of Europe during the Middle Ages (600-800 kilograms per hectare [kg/ha]). This yield was adequate prior to 1960, because the population had stayed within the bounds necessary to ensure adequate food supplies even with suboptimal agricultural practices.

In the 1960s, however, India and China were experiencing improved living conditions and modern health-care techniques that prolonged life expectancy, which, compounded with high fertility, could have led to massive malnutrition. Fortunately, at that time, rice production increased substantially throughout much of East and Southeast Asia, largely from strains developed at the International Rice Research Institute (IRRI) in the Philippines. Founded in 1960 by the Ford and Rockefeller Foundations, IRRI began its research activities in 1962 with the development of semi-dwarf breeding lines for rice.

Subsequently, rice production throughout East and Southeast Asia grew from 52 million metric tons in 1961 to about 125 million metric tons in 1991 a much faster growth rate than that of the regional population, which had almost doubled over that time.When India gained independence from Britain in the late 1940s, Pakistan was partitioned into an independent nation, causing India to lose major areas of irrigated wheat land in the west, vast riceproducing areas in the east, and important agricultural research and education facilities.

The country was facing a crisis. N. C. Mehta, secretary of the Indian Council of Agricultural Research, stated in 1951, "The country can rise as a whole only if our agricultural economy with its millions of farms and lakhs of villages is to revive with a new sense of energy and well-being". In the two decades that followed the partitioning, India tried several methods to deal with potential food shortages: expanding the amount of farmland, establishing price controls, launching community development programmes, and obtaining grants and low-cost sales of surplus crops from the United States and other nations.

Unfortunately, other government policies-import substitution, cheap food, and industrialisation at the expense of agriculture-rendered these efforts largely ineffective. After some initial hesitancy, parts of India and other South Asian nations finally embraced Green Revolution technologies. Those measures paid off. Rice production more than doubled from 1961 to 1991, and wheat production almost quadrupled in that time, more than matching the approximate doubling of the population in South Asia. Thus, steady and significant increases in crop production throughout East, Southeast, and South Asia, as in Latin America, prevented the malnutrition that might have otherwise occurred with the soaring populations of the second half of the 20th century.

The United Kingdom provides an example of the Green Revolution's significant impact on the industrial world. In the post-Colonial era from 1935 to 1954, Britain was concerned over whether its lands would produce sufficient food for its population because it had previously depended on imports from the colonies. It was especially concerned about wartime food production in the early 1940s. Hence, throughout the United Kingdom, growers adopted the Green Revolution technologies, particularly the high-yield crop varieties. The adoption of these varieties paid off: Britain produced only 23 percent of its wheat consumption in 1936-1939, but had boosted that rate to 67 percent in 1974-1975 and to 77 percent in 1980- 1981. Whether or not an actual food crisis would have occurred without the Green Revolution technology, Britain did become more independent with regard to its key food crops.

Despite its successes in many parts of the world, the Green Revolution has not yet made a significant impact on most areas of Africa. In the later part of the Green Revolution, four CGIAR centers were founded in Africa. These centers have contributed to increases in the production of maize, bananas, cassava, and rice. But these increases have been limited, and the growing population has effectively neutralised these gains. Overall, while average agricultural production in Asia after the Green Revolution has increased dramatically to nearly three tonnes per hectare from 1960 to 1990, Africa's production level has fallen to about one tonne per hectare, which is about the average productivity of British farmers during the reign of the Roman Empire.

One possibility is that sub-Saharan Africa features types of agriculture that could not benefit from the technological packages developed through the Green Revolution for use in Asia and Latin America. Therefore, the criterion of substantial production increases could not be met. The new HYV crops that were developed elsewhere in the world were not suited to African planting conditions, where the topsoil is thinner and weather patterns, such as periods of drought, are more unpredictable. Serious and uniquely African agricultural problems persist. The staple crop-maize-is attacked by insects, streak virus, and the parasitic weed Striga that extracts nutrients from roots.

Mealy bugs and mosaic virus attack cassava. Weevils, nematodes, and the fungal disease black Sigatoka attack bananas. Fungal diseases shrivel and weaken bean pods, and drought is regularly occurring. These problems would have been difficult to over-come through even the enhanced conventional means that the Green Revolution employed. A number of the CGIAR centers, such as the International Institute of Tropical Agriculture (IITA) and the Centro Internacional de Agricultura Tropical, or International Center for Tropical Agriculture (CIAT), have developed improved varieties of sorghum, millet, cowpeas, cassava, potatoes, and sweet potatoes for African use.

But the sorts of dramatic yield improvements scored with rice and wheat in Asia have not been achieved in Africa. Also, infrastructural problems have continued to serve as a barrier to Green Revolution success in Africa. Inadequate property rights protection have had the effect of blunting farmers' production opportunities and incentives, leaving little possibility for private gain in adopting new technologies. Local banks have been unable or unwilling to assist in providing the necessary loans for farmers to purchase new technologies.

It is also possible that cultural constraints that control particular divisions of labor among family members or particular patterns and timing of planting, cultivating, and harvesting may be stronger in Africa than in Asia and Latin America. Even if there were a greater degree of cooperation among the various African institutions in promoting Green Revolution practices, and even if agricultural technologies would have proven to be more successful, geography alone would limit the Green Revolution's success in Africa. In the 1700s, Adam Smith, in his Wealth of Nations, wrote that the inland parts of Africa would lag in economic development because of problems with transportation of goods.

Moreover, geography may affect other important determinants of agricultural development, such as communications, human and animal health, natural resources, and population density, as well as transportation. The African transportation system was poorly prepared to deliver Green Revolution technologies to the places they were needed. Uganda and Ethiopia, for example, had fewer than 100 kilometers of paved roads per one million people as of 2001-roughly 100 times fewer kilometers of roadway than in other developing nations such as Brazil and India, and some thousand times less than in industrialised nations such as the United States and France.

In addition, many of the roads in Africa were designed to lead to mines rather than farmlands. Because of these transportation obstacles, Green Revolution technologies could not reach African farmers easily or reliably. Partly as a result of these limitations, malnutrition has persisted and indeed grown even worse in sub-Saharan Africa, unlike in other regions of the world where malnutrition is on the wane. Through the efforts of the CGIAR centers that were established in Africa and other important measures, agricultural development eventually began with specific crops that are of particular importance to the diets of sub-Saharan Africa.

It has been argued that Green Revolution efforts came to fruition in a particular context characterised by fear of famine, overpopulation, and the threatened rise of communist governments in areas considered a strategic threat to the West. If so, then motivations for the Green Revolution reach far beyond the agricultural and philanthropic realm. A number of various political motivations likely shaped the success of the Green Revolution in certain parts of the world, demonstrating on a larger scale the importance of cooperation among the various stakeholders to enable the agricultural revolution to succeed. Governmental agendas, both domestic and international, were altered to accommodate the new agricultural advances.

In the mid-20th century, it was unquestioned that greater food production would lead to greater political stability globally as well as to greater prosperity and security in developing nations. Various nations at various socioeconomic levels all saw the need for vastly increased agricultural production, for somewhat different reasons. In addressing the situation in Mexico, for example, Perkins argued that beyond the desire to produce more food, the Mexican Agricultural Programme's primary proponents within Mexico wished to completely reshape the Mexican economy and intended to use Green Revolution technologies as a means to that end.

Namely, to create a modern industrial state, agriculture would need to make use of new technologies that were more labor-efficient so that a large number of farmers could choose (or would be forced) to leave the rural areas and become part of the industrial workforce. Former human labor from agriculture would have to provide at least part of the capital for Mexican industrialisation. India may have embraced the Green Revolution partly to recover from the damages caused by British imperialism and the shattering of the economy of British India after its independence.

The security and autonomy of the Indian nation and foreign exchange considerations were the prime drivers in the national commitment to support crop breeding. Although Gandhi had envisioned an India devoted almost entirely to agriculture, many Indian leaders wished to develop the nation's industrial sector for the purpose of remaining competitive in the world market. They decided that if industrialisation were to occur, the wealth of the countryside had to finance most of that industrial development · hence, the need for improved food yield. Moreover, the major impetus for the rapid uptake of HYV seeds in South Asia was the two consecutive crop failures during the monsoons of 1966 and 1967.

The United States, which was the only country in the world carrying food reserves, shipped much of its surplus grain to India during those years; however, such dependence on foreign aid was recognised by both India and the United States as being risky and undesirable. The UK, too, expanded its commitment to crop breeding as it struggled to reconstruct its post-imperialist economy. After the loss of its significant food-producing colonial lands worldwide, the British government saw the need for an independent food supply as an important component of its political agenda in the mid-20th century. Prior to that time, Britain's only prosperous area of agriculture was livestock production. However, the emergency times of World War II increased the need for British food production of crops, and the loss of India, in particular, as a food producer in the late 1940s further brought about policy pressures to improve agricultural production through Green Revolution methods.

Much of the development aid worldwide has supported large defense-oriented or commercial projects, with the improvement of living standards as a secondary goal at best. A case in point concerns aid given for the Pergau Dam project in Malysia. The Overseas Development Administration in the United Kingdom had advised strongly against giving aid to this project because of its low developmental value in the region, but its opinion was overruled on the grounds of safeguarding British defense contracts. Although this is a different situation from the agricultural pursuits in developing nations, similar national security concerns influenced the decision to provide aid for the agricultural technology transfer.

Although the Green Revolution did in fact significantly reduce hunger and poverty in many parts of the developing world, the international motivation for its success extended beyond the goal of poverty reduction. In particular, the United States made significant commitments at the federal and philanthropic levels to promote crop breeding as part of the Cold War effort to contain the spread of communism and possibly to foster other economic and foreign policy objectives.

Challenges of Green Revolution

While the Green Revolution had a dramatic and lasting effect for the better, it did not solve agricultural problems worldwide.

Agricultural Challenges

The effectiveness of the Green Revolution is reaching a natural limit, even as human populations in parts of the developing world continue to rise. The population of the industrialised world may increase by only 4 percent from 2000 to 2020, reaching about 1.24 billion. The population of Africa, on the other hand, may well increase by 50 percent in that time and the population of the developing world as a whole by 36 percent. By the year 2030, it is estimated that more than eight billion persons may populate the world. It is noteworthy that Africa, due to the lagging success of the Green Revolution on that continent and its greatest share of the population increase, is most in need of a new agricultural revolution.

Studies have shown enormous productivity gains in the early decades of the Green Revolution, followed by slower gains more recently. We can see that although agricultural growth has continued to be positive in the past decade, the rate of growth has slowed dramatically. Aside from technological limitations, farmland limitations make future increases in the food supply difficult to achieve.

Many agricultural sectors of Africa and Asia have no suitable untilled land left. In 1961, there were 0.44 hectares of farmland for every person in the world. In 2000, the number fell to 0.26 hectares, and projections indicate that by 2050 the number will fall to 0.15. Thus, from the standpoint of available arable land, the situation does not look promising for the future, if the technological status quo continues as it has. To increase production, it is necessary to improve yields on currently existing fields, beyond what Green Revolution technologies have been able to provide.

Health Challenges

Although the Green Revolution did stave off hunger to a significant extent on two continents, an estimated 800 million people still do not have access to sufficient food to meet their needs, which will lead to malnutrition. Chronically undernourished people have an insufficient caloric intake to meet their daily energy needs, and malnutrition also results from lack of proteins, vitamins, minerals, and other micronutrients in the diet. The most important sources of calories are cereals, the types of crops for which plant breeders developed HYVs during the Green Revolution; cereal diets that supply sufficient calories usually provide enough protein as well.

Malnutrition can lead to death, not just through starvation, but because a poor diet leaves humans especially vulnerable to illnesses such as diarrhea, measles, respiratory infections, and malaria. Children whose mothers were malnourished and are themselves lacking basic nutrients and calories are at further risk of a number of developmental disorders. The United Nations Children's Fund (UNICEF) estimated that malnutrition is the cause of about 12 million deaths annually of children under the age of five in developing countries. Aside from not having enough to eat in terms of calories, people in developing nations often eat unbalanced diets. For example, those who eat primarily boiled rice or maize porridge rarely consume enough meat or green vegetables to balance their nutritional needs.

Meat, vegetables, fruit, and fishery products are important sources of vitamins and micronutrients that are lacking in cereals. Developed nations have vitamin pills and enriched foods to make up for potential dietary shortages, but pills and enriched food are not yet practical options for the developing world. Specifically, lack of proper nutri-tional variety in African diets, in some cases along with inadequate calories, has resulted in 65 percent of African women of childbearing age being anemic, 40 million children being severely underweight, and 50 million being vitamin-A deficient. Although food prices have become lower and many developing nations produce enough food in the aggregate, many people still go hungry.

Indeed, food intake among the poorest people experienced negligible improvement throughout the Green Revolution. This is particularly true in poor rural areas where food prices are still too high, and people cannot grow enough food or buy enough food because their income is too low or they are unable to find food in other ways. A large number of the poorest and most oppressed people are women.

Environmental Challenges

While the Green Revolution played a key role in achieving food security and reducing rural poverty in key areas of the developing world, it was environmentally harmful in many settings. Some of the bestirrigated lands have become overly salinesince irrigation was introduced as part of Green Revolution technology, which has made crop planting difficult if not impossible. Irrigation cannot continue today in many areas because of decreasing water availability and resulting soil salinity. These conditions were exacerbated during the Green Revolution by the need for irrigated lands so that high-yield varieties could succeed.

In addition, water tables have diminished, leading to problems with water accessibility for agriculture. Moreover, waterways and soils have become contaminated by the large amounts of pesticides and fertilisers used on Green Revolution farmlands. Pesticides and nitrates in drinking water have proven to be detrimental to human and animal health. An unfortunate consequence of overuse of pesticides in particular areas is that crop pests have developed resistance to the pesticidal chemicals, rendering the chemicals ineffective. Indeed, these environmental spillovers have had a depressing effect on agricultural production, the very thing they were intended to improve.

5

Bioengineered Crops

The impacts of genetically engineered (GE) crops on farm profitability vary greatly by region, crop, and technology. Impacts also vary with seed premiums, crop prices, and prices of alternative pest control programmes. Moreover, some factors that influence adoption of GE crops are difficult to measure (for example, the economies in management time associated with the adoption of herbicide-tolerant crops). Finally, profits may be affected by factors other than GE adoption, such as other cropping practices, weather, or management ability, making it difficult to isolate the effect of GE crop varieties.

Producers of herbicide-tolerant crops versus traditional crops benefit mainly from lower costs. They expect to achieve at least the same output while lowering weed control costs for chemicals, chemical applications, mechanical tillage, and scouting. In return, producers pay more to seed companies for the herbicide-tolerant seed. Thus, the profitability of the herbicide-tolerant programme depends on weed control cost savings compared with seed cost premiums. Seed companies aim to set the seed price high enough to obtain as much of the farmers' savings in weed control costs as possible, while still inducing the producer to use the herbicide-tolerant seed. In addition, the substitution of glyphosate, used in most herbicide-tolerant programmes, for other herbicides decreases the demand for those herbicides. Thus, the prices of other herbicides decrease, lowering production costs even for those farmers not using the herbicide-tolerant crops.

Other factors believed to affect the economics of adoption of herbicide-tolerant crops are the simplicity and flexibility of the weed control programme. Herbicidetolerant programmes allow growers to use one product instead of several herbicides to control a wide range of both broadleaf and grass weeds without sustaining crop injury. Thus, herbicide-tolerant crops appear to free up valuable management time for other activities. However, standard measures of net returns to management have not been designed to quantify how management intensive a technology is in dollar terms. Potential users of Bt crops (Bt corn or cotton) face a complex decision in determining the relative profitability of these technologies.

The use of Bt seed can reduce costs by virtually eliminating the application of insecticides intended to control Bt target pests. More important, because chemical insecticides are not as effective as the control achieved with Bt seed, planting Bt seed increases crop yields, as crop losses are reduced. Therefore, Bt crops are more profitable than traditional insect control measures only if the target pest infestations are severe enough to cause economic losses greater than the economic impact of the price premium paid for the Bt seed. However, unlike annual weed infestations that are relatively stable and predictable, insect infestations can vary dramatically each year. Since the decision to plant Bt crops must be made prior to observing the insect infestation, the farmer may or may not make the most economical decision for a given year depending upon the resulting infestation. Thus, Bt crops act as insurance against significant losses that may occur in the event of severe pest infestations.

Consumers express their preferences for bioengineered crops at the market and producers must respond to the economic signals that these preferences convey. Factors influencing consumer preferences include

(1) their perceptions of benefits and risks of bioengineered crops on human health and the environment,

(2) their ethical stance toward genetic engineering, and

(3) their trust in government regulations concerning risk assessment and management.

The importance of these factors has varied substantially among consumers both within and among countries, causing significant uncertainty about the acceptance of bioengineered crops, particularly in international markets. This uncertainty may discourage adoption of these crops, particularly food crops. In addition, specific markets for nonbiotech crops have emerged as consumers have expressed their preferences.

Environment Benefits

While many of the environmental benefits and risks of GE crop adoption are dificult to quantify, changes in pesticide use associated with the adoption of GE crops are surely an important effect of GE crops. Several recent polls among consumers indicate that consumers were willing to accept biotechnology as a means of reducing chemical pesticides used in food production. More specifically, consumers would be likely to buy a variety of produce "if it had been modified by biotechnology to be protected from insect damage and required fewer pesticide applications".

While profitability (i.e., the extent of yield increases and/or input cost reduction versus the costs of adoption relative to the current management practices) is key to explaining the extent and rate of technology adoption, most studies acknowledge that heterogeneity among farms and farm operators can often explain why all farmers may not adopt an innovation in the short or long run. Differences influencing readiness to adopt include farm size, tenure, operator education/experience, and access to information and credit. The nature of the technology or the financial, locational, and physical attributes of the farm may also influence profitability and, ultimately, the adoption decision.

Other factors that may have some effect on adoption include the interaction of GE crops with other cropping practices. For example, the adoption of herbicidetolerant crops complements the conservation tillage practices and narrow row spacing. Adoption may also have some impact on the safety of farmworkers and other people operating (or living) in nearby areas. For example, as the use of Bt crops ensures that insect control is properly timed and reduces the need to handle and apply synthetic insecticides, it thereby increases farmworker safety and avoids the misapplication or drift of chemicals from the target area.

Diffusion of Bioengineered Crops

Diffusion is the process by which a successful innovation gradually becomes broadly used through adoption by firms or individuals. Many agricultural innovations follow a well-known diffusion process: after a slow start in which only a few farmers adopt the innovation, the extent of adoption (the fraction of potential users that adopted the innovation) expands at an increasing rate. Eventually, the rate of adoption tapers off as the number of adopters begins to exceed the number of farmers who have not yet adopted.

Finally, adoption approaches asymptotically its maximum level, until the process ends. This process generally results in an S-shaped diffusion curve, first discussed by rural sociologists and introduced to economics by Griliches in 1957. Two types of diffusion models-static and dynamic-have been used to examine the progress of agricultural innovations. Static diffusion models, following the terminology of Knudson, are those growth models that represent the adoption path, expressing the percentage of adopters only as a function of time (they do not contain any other factors).

Two characteristics of static models suggest their unsuitability to model some innovations. First, they have a predefined point of maximum adoption as a share of the total population. Second, adoption must always increase over time until it reaches this maximum. Unlike static diffusion models, dynamic diffusion models allow the coefficients (fixed in static models) that determine the diffusion path to be functions of economic or other factors that affect diffusion. Moreover, dynamic diffusion methods relax some of the assumptions of static diffusion models by allowing for disadoption, and help directly identify and measure the impact of variables significant to the adoption of an innovation.

Diffusion of Genetically Engineered Crops

The diffusion of genetically engineered (GE) crops appears to have followed an S-shaped diffusion curve in 1996-99, and the static logistic model appears to fit the data. However, the market environment during the past few years, particularly the export market, suggests that use of static diffusion methods may be inappropriate to examine the diffusion of this technology. Increased concern, especially in Europe and Japan, regarding the safety of GE crops has resulted in the development of segregated markets for nonengineered crops. While these markets are still small, the 2000 data regarding the adoption of these crops suggests that dynamic considerations may be necessary to examine this particular adoption process.

The diffusion of GE crops is modeled by specifying a variable-slope logistic function. Following Griliches, the variable rate of acceptance (slope) is modeled as largely a demand, or "acceptance," variable. The model is estimated using adoption data obtained from the following USDA surveys: the ARMS surveys for 1996-98 data, the NASS Crop Production survey for 1999, and the NASS Acreage survey for 2000. The crops included in the surveys are corn, soybeans, and upland cotton. Prior to model estimation, it is necessary to specify the ceilings, or maximum adoption levels, of different genetically engineered crops. These ceilings are based on limitations due to farm production considerations or market restrictions. That is, for many technologies, not all farmers are expected to adopt the technology.

The base-case ceilings for Bt crops are computed by considering infestation levels and refuge requirements. For example, Bt crops would likely not be adopted on acreage where pest infestation levels do not exceed the economic threshold for treatment. In the case of herbicide-tolerant crops, a ceiling computed from weed infestation levels is not likely to be binding, since most acreage is potentially susceptible to infestation. For this reason, ceilings in these cases are based on other considerations. For the diffusion of herbicide-tolerant soybeans, the ceilings are computed based on demand considerations arising in the export market.

Bt corn is relatively sensitive to the scenario (ceiling) specification. A 30-percent higher corn-borer-infestation scenario projects a Bt corn adoption level (for 2001 corn acreage) 15 percent above the base-case projection; the 30-percent lower infestation projects a level 32 percent below the base-case projection. In contrast, the comparable numbers for Bt cotton are 4 percent and 3 percent, respectively. With no export restrictions, the projected adoption rate for herbicidetolerant soybeans is 18 percent above the base-case projection (no GE exports).

For herbicide-tolerant cotton, the 70-percent adoption ceiling scenario projects an adoption rate of 15 percent below the base-case (90-percent ceiling) projection. Since technological change can affect the level of output, product quality, employment, trade, real wages and profits, the adoption of new technologies offers economic opportunities and challenges. Consequently, understanding the adoption process continues to be of interest to economists, sociologists, and policymakers.

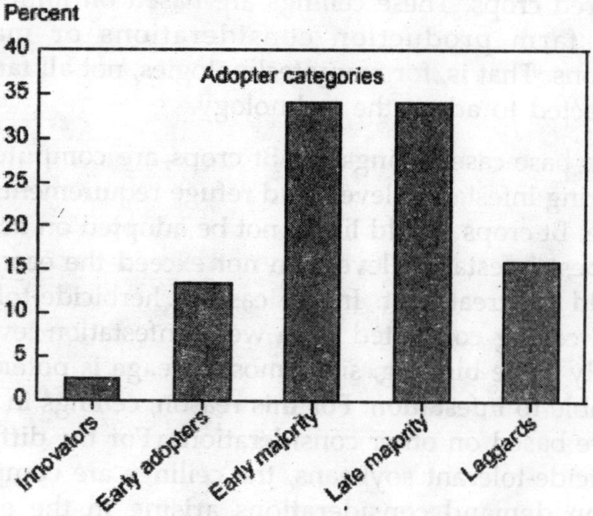

Figure 1. Rogers' characterisation of adopters

Of particular interest to policymakers is the impact of new technologies on farm structure (i.e., farm size), and the role that farm structure plays in the adoption process. Numerous technology adoption studies have been conducted over the last 40 years, beginning with Griliches and Rogers. Feder, Just, and Zilberman and Feder and Umali review many of these studies. Of note, Rogers hypothesised that innovators or early adopters (figure 1) have attributes different from later adopters or nonadopters.

Influence of Size of the Farm

Characteristics of GE crop technologies led us to expect that adoption would be invariant to farm size. GE crop technologies are embodied in the variable inputs (e.g., seeds), which are completely divisible. Thus, GE crops may be used in any amounts, unlike technologies embodied in "lumpy" inputs like tractors or other machinery, which require extensive capital investments and many acres over which the operator can spread the costs of acquisition. However, actual mean adoption rates for different farm sizes, obtained directly from 1998 USDA survey data, are not constant.

Adoption rates appear to increase with size of operation for all the technologies, but in different patterns. For example, the adoption of herbicide-tolerant soybeans and corn was fairly stable (39-52 percent for soybeans and 6-10 percent for corn) for farms above 50 acres. In contrast, the adoption of Bt corn increased continuously with the size of the operation.

Table 1. Factors affecting the adoption of GE crops

Factor	Herbicide-tolerant soybeans	Bt corn	Herbicide tolerant corn
		Elasticity	
Education	0	0.179	0.336
Experience	0.236	0	0.453
Marginal region	-0.079	0	0

Size	0	0.258	0.279
Risk	-0.859	0	0
Limited-resource	-0.049	0	0
Contract	0.036	0.022	0
High infestation	na	0.123	na

While illustrative, this comparison of means would be valid only in an ideal experimental setting where factors other than size are "controlled" by making them as similar as possible. Thus, differences in mean adoption rates cannot necessarily be attributed to size since survey results are influenced by many other factors not controlled for, including location, access to credit, risk, wealth, other cropping practices, etc.

However, the adoption of herbicide-tolerant corn and Bt corn are positively related to farm size. The different empirical results obtained for the adoption of herbicidetolerant soybeans (invariance to farm size) and herbicidetolerant corn (adoption positively related to farm size) from the Tobit analysis may be understood by examining their adoption rates. The 1998 adoption rate for herbicide-tolerant soybeans in the sample (34 percent of farms) implies that adoption of herbicide-tolerant soybeans has progressed past innovator and early adopter stages (fig. 1) into the realm where adopting farmers are much like the majority of farmers.

On the other hand, the adoption of herbicide-tolerant corn was quite low in 1998 (5 percent of farms), implying that adoption of this technique was largely confined to innovators and other early adopters who tend to control substantial resources and are willing to take the risks. This result is consistent with Rogers' observation that adoption is more responsive to farm size at the innovator stage and that the effect of farm size on adoption generally diminishes as diffusion increases. This effect can be more closely examined by using the decomposition of the responsiveness (measured by the elasticity) of adoption with respect to size into components that reflect the behavior of users and nonusers of the technology. The decomposition of the elasticity of size for each technology reveals that adoption of herbicide- tolerant corn by nonusers was much more responsive to changes in size than for the other technologies (0.442 percent per 1-percent increase in size). The results for Bt corn are more difficult to interpret.

Bt corn was adopted by 20 percent of farms in 1998, a level inclusive of more than innovators and other early adopters. However, unlike herbicidetolerant soybeans, the estimated adoption elasticity of Bt corn with respect to size was positive and significant. One important difference between Bt corn and the herbicide-tolerant technologies is that Bt corn is designed to target a pest with much more spatial variation than pests targeted by the other technologies. European corn borer (ECB) infestations are quite severe in some areas and virtually nonexistent in others.

In fact, many western cornproducing States (e.g. Iowa, Nebraska) tend to have large corn farms and are also more likely to have high ECB infestations. It would have been interesting to examine if Bt corn adoption increases with size among farms with an ECB infestation above a certain threshold, but this analysis could not be conducted because of insufficient data. Moreover, the responsiveness of the adoption to farm size was largest for the farms that had already adopted Bt corn in some of their corn acreage.

Adopters of Bt corn may. choose to plant Bt corn throughout the operation because ECB infestations tend to be widespread. Thus, once the decision to control for ECB is made, it is implemented across much of the operation. However, the probability that a given farm would have all their acreage under adoption was lower for Bt corn than herbicide-tolerant adopters because of refuge requirements associated with Bt corn. To conclude, the interrelationships between the attributes of innovations and the characteristics of adopters at different stages of the diffusion process make it difficult to examine the influence of size on the adoption of innovations.

Theoretically, this difficulty is inherent to the data commonly available and could be surmounted if the model included every factor that characterises an innovator such that size effect could be totally isolated. In practice, one uses proxies for most relevant factors, such as risk, operator management skills, etc., which are limited by the quality of the data. It is doubtful that adoption patterns can be used to categorise technologies as size invariant or size dependent because such categorisation depends not only on the attributes of the innovation but also on the extent of adoption and characteristics of the adopters. Moreover, technology adoption is a dynamic process.

Influence of Other Factors

Adoption of all three of the GE crop technologies was positively and significantly influenced by operator's education, experience, or both. These factors may also reflect management quality in the sense that more educated or experienced operators are more likely to understand that the economic benefits of new technologies usually accrue to early adopters. The use of contracting (marketing or production) was positively associated with technology adoption in most of the models. The effect of contracting may be indicative of the greater importance placed on risk manage-ment by adopting farms.

Contracting locks in a commodity price or service fee and ensures a market for GE crops, lessening price and any market access risk that could result from uncertain consumer acceptance of these crops. GE crops also reduce production risks by lowering the likelihood of yield losses due to weed and insect pressure. Location of the operation outside of the primary production area was associated with a lower expected adoption for herbicide-tolerant soybeans. Among the farm typology variables, only the limited-resource classification variable was significant in the model for herbicidetolerant soybeans, and these farms were less likely to adopt.

Corn borer infestation had a significant and positive influence on the expected adoption of Bt corn. Credit reserves, off-farm work, and land tenure were not significant in any of the adoption models. Compared with other agricultural innovations, such as soil conservation improvements, these factors are probably less important to the adoption of GE crops because of their unique attributes, such as low fixed costs.

Impact of Adopting Bioengineered Crops

Faced with reduced returns to crop production caused by low commodity prices, farmers are examining alternative technologies as ways to improve financial performance by cutting costs and/or increasing yields. Rapid adoption of GE crop varieties among farmers suggests that these technologies are perceived to have advantages over traditional methods. GE crop varieties with pest management traits provide a broad spectrum of potential benefits and appeal to farmers because they promise to simplify pest management, reduce its costs, increase its effectiveness, and increase flexibility in field operations.

But impacts vary by crop and technology and are often confounded with other factors, making it difficult to isolate the effect of adopting GE crop varieties on yield and profits. For example, the physical environment of the farm (e.g., weather, soil type) affects profitability directly through increased fertility and indirectly through its influence on pests. Field-level data provide more accurate information regarding yields and input uses; whole-farm data allows the calculation of broader measures of farm financial performance.

Econometric Models

Field-level analysis

The field-level analysis used the econometric model developed by Fernandez-Cornejo, Klotz-Ingram, and Jans to estimate the impact of adopting GE crops on yields and net returns using 1997 field-level survey data. The model takes into consideration that farmers' adoption and pesticide use decisions may be simultaneous. In addition, the model corrects for self-selection. Finally, the model is consistent with farmers' desire to maximise profits. The field-level analyses include herbicide-tolerant soybeans, herbicide-tolerant cotton, and Bt cotton.

Each technology was modeled individually using 1997 survey data. Each model included pest infestation levels, other pest management practices, crop rotations, and tillage. Geographic location was included as a proxy for soil, climate, and other local factors that might influence the impacts of adoption. Net returns are defined as per-acre revenues minus peracre variable expenses, including pesticides, seed (including technical fee), and labor.

Whole-farm analysis

The whole-farm analysis assesses the impact of adopting GE crops on farm financial performance using the econometric model. By controlling for several other factors that may also affect financial performance-such as economic and environmental conditions, management practices, and operator characteristics-the model attempts to isolate the effect of GE crop adoption on farm financial performance. Separate models were estimated for herbicide-tolerant corn, Bt corn, and herbicide-tolerant soybeans. The models were specified using variables that have shown to be related to technology choice in previous research.

Several measures of farm financial performance were examined, but results are reported for only two measures: net returns per tillable acre and modified net farm income per tillable acre. Net returns were measured as gross value of crop production minus total farm chemical and seed expenses, where gross value of crop production is the production of each crop commodity produced on the farm operation valued at the State-average price received by farmers. This measure of financial performance was used because most of the financial impacts of adopting GE crops result from changed crop yields, reduced chemical costs, and increased seed costs. Thus, this measure captures the greatest influence that GE crop adoption would have on whole-farm financial performance as it filters out the impact that other farm activities -such as livestock production, custom work, and government programme participation- have on financial performance.

Moreover, this measure is consistent with the "net returns" variable used in the field-level analysis as well as other studies on the relative economies of GE and conventional crops. Modified net farm income was measured as net farm income (NFI) plus interest expense. NFI was calculated as gross farm income minus total farm operating expenses (excluding marketing expenses). The measure of net farm income used in this analysis measures the return to operator and unpaid family labor, management, and capital (both equity and borrowed). Interest expenses are added back to net farm income so that variation in farm debt does not influence the financial comparison among farms. Because of the influence of several factors on MNFI, the impact of GE crop adoption would need to be relatively strong in order to have a significant effect on MNFI.

The whole-farm analysis of the impact of adopting GE corn (soybeans) was conducted on two segments of the farm population: (1) operations that harvested one or more acres of corn (or soybeans), and (2) operations that specialised in the production of corn (or soybeans), with more than 50 percent of the total value of farm production from corn (soybeans). Such specialised farms were examined in addition to all growers because GE technologies likely have the greatest financial impact on operations specialising in the target commodities. The whole-farm analysis also examined the effect of spatial variation on the impact of GE crop adoption using the ERS farm resource regions.

Because pest infestations differ across the country, one would expect the impacts of pest control measures such as GE crops to be greatest where target pest pressures are most severe. Research suggests that the value of Bt corn relative to conventional varieties increases as one moves from east to west in the Corn Belt, because ECB infestations are much more frequent and severe in the western Corn Belt. Also, weed pressure tends to be greatest in the eastern and southern United States because of the hot, moist climate and the longer growing season. Therefore, the expected returns of herbicide-tolerant crops would be greater in these areas because of higher costs for conventional weed control.

Empirical Results

Field-Level Results

GE crops available for commercial use do not increase the yield potential of a variety. In fact, yield may even decrease if the varieties used to carry the herbicidetolerant or insect-resistant genes are not the highestyielding cultivars. However, by protecting the plant from certain pests, GE crops can prevent yield losses compared with non-GE varieties, particularly when infestation of susceptible pests occurs. This effect is particularly important in Bt crops. For example, before the commercial introduction of Bt corn in 1996, the European corn borer (ECB) was only partially controlled using chemical insecticides.

The economics of chemical use were not always favorable and timely application was difficult, so farmers often accepted yield losses (of 3 to 6 percent per one corn borer per plant depending on the stage of plant development) rather than incur the expense. From this perspective, for the cases analysed, the empirical results are not surprising. Adoption of herbicide-tolerant soybeans led to a small but statistically significant increase in yield while adoption of Bt cotton led to a large increase in yields. A 10-percent increase in the adoption of herbicide-tolerant soybeans led to a 0.3-percent increase in yields. On the other hand, an increase of 10 percent in the adoption of adoption of Bt cotton in the Southeast increased yields by 2.1 percent.

The adoption of herbicide-tolerant cotton also has a positive and statistically significant effect on net returns (elasticity is +0.18), as does the adoption of Bt cotton (elasticity of +0.22). However, the adoption of herbicide-tolerant soybeans does not have a statistically significant effect on net returns. The soybean results appear to be inconsistent with the rapid adoption of this technology. Yet, other factors have a considerable impact on adoption, namely the simplicity and flexibility of the weed control programme, which frees up valuable management time for other activities. However, it is difficult to measure management involvement on various technologies from survey data.

Whole-Farm Results

GE crop adoption was found to affect net returns on specialised corn farms. Adoption of herbicide-tolerant corn had a positive and statistically significant effect on net returns, but the elasticity of net returns with respect to adoption was negative for Bt corn. The effect of GE crop adoption on farm financial performance was not significant for soybean farms. An analysis using broader financial performance measures (including net farm income and return on assets) did not show GE crops to have a significant impact. GE crop technologies do not require a capital investment and, thus, their impact on farm finances is mainly limited to changes in variable costs and returns. For this reason, adoption-impact models are likely to be more useful in explaining net returns than in explaining farm income.

The impact of GE crops on the net returns of specialised corn farms varied by region. On all specialised corn farms nationwide, a 10-percent increase in herbicide-tolerant corn led a 2.7-percent rise in net returns. But in the eastern Heartland, the increase in net returns expanded to 4.1 percent, consistent with high weed pressures there. In contrast, the adoption of Bt corn led to a decrease in net returns among specialised corn farms; as adoption increased by 10 percent, returns declined by 3.4 percent. This effect was much less in the western Heartland than the eastern (elasticity of -0.27 versus -0.46). Corn borer pressure is greater in portions of the western Heartland, as are the benefits of its relief.

Perhaps the biggest issue raised by these results is how to explain the rapid adoption of GE crops when farm financial impacts appear to be mixed or even negative. Both herbicide-tolerant cotton and Bt cotton showed positive economic results, so rapid growth in adoption is not surprising in these cases. The financial benefits of adopting herbicide-tolerant corn may be due in part to seed companies setting low premiums (including technology fees) relative to conventional corn varieties in an attempt to expand market share. Also, the limited acreage on which herbicide-tolerant corn has been used is likely acreage with the greatest comparative advantage for this technology, boosting its financial benefits.

A primary motivation may be the simplicity and flexibility of the herbicide-tolerant programme, which allows growers to use one product instead of several herbicides to control a wide range of both broadleaf and grass weeds, and also makes harvest "easier and faster". Herbicide-tolerant crops also fit into ongoing trends toward postemergence weed control, conservation tillage practices, and narrow row spacing. In addition, the window of application for glyphosate is wider than for other postemergence herbicides, allowing growers to treat later with less concern about getting poor weed control or injuring the crop. Because glyphosate has no residual activity, carryover restrictions are not a problem, giving growers more rotation options.

Glyphosate is also effective at controlling weeds that are resistant to other classes of herbicides. The economic potential of Bt corn on an individual farm is more difficult to evaluate because returns to Bt corn are realised only if the density of European corn borer (ECB) is sufficient to cause economic losses greater than the premium paid for the Bt seed. This requires farmers to forecast infestation levels and input and corn prices before planting, prior to observing an infestation. By one account, only 25 percent of corn acreage was infested with ECB at a treatable level in 1997. This would conform with Bt corn adoption rates of 19 percent of the corn acreage in 1998 and 26 percent in 1999.

Use of Pesticide

Herbicides constitute the largest pesticide category in U.S. agriculture. Corn is the largest herbicide user, and 96 percent of the 62.2 million acres devoted to corn production in the 10 major corn-producing States were treated with more than 164 million pounds of herbicides in 1997. Soybean production in the United States also uses a large amount of herbicides-97 percent of the 66.2 million soybean acres in the 19 major producing States were treated with more than 78 million pounds of herbicides in 1997.

Cotton production relies heavily upon herbicides to control weeds, often requiring applications of two or more herbicides at planting and postemergence herbicides later in the season. Close to 28 million pounds of herbicides were applied to 97 percent of the 13 million acres devoted to upland cotton production in the 12 major cotton-producing States in 1997. Cotton production also uses a large amount of insecticides, with 77 percent of upland cotton acres (in the 12 major producing States) treated with 18 million pounds of insecticides in 1997.

While only 30 percent of the corn acres received insecticides in the 10 major corn-producing States in 1997, the amount of these insecticides exceeded 13 million pounds. Pesticide use on corn and soybeans has declined since the introduction of GE corn and soybeans in 1996 (fig. 2). Field tests and enterprise studies have analysed the agronomic, environmental, and economic effects of adopting GE crops, including actual pesticide use changes associated with using GE crops.

Many of these studies have suggested that insecticide use has (or will) decline with the adoption of Bt varieties and that herbicide use is reduced with herbicide-tolerant varieties. The use of Bt varieties has led to reductions in those insecticides previously used to treat the pests targeted by Bt.

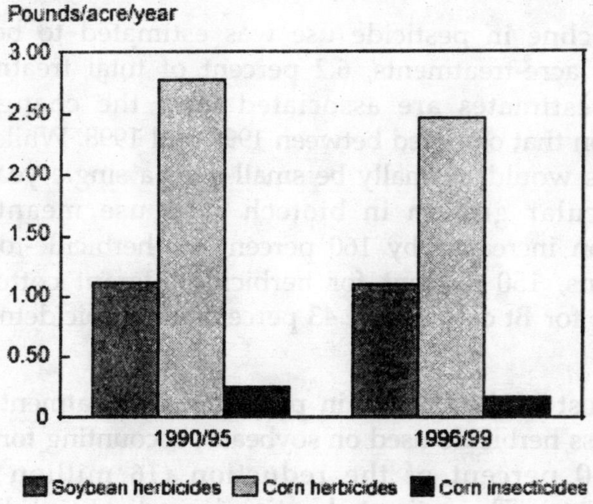

Figure 2. Pesticide use, 1990-99

However, conventional insecticides targeting insects not affected by the toxin continue apace. With herbicide-tolerant crops, which facilitate the use of a particular herbicide such as glyphosate, adoption simply involves substitution, changing the mix of herbicides used in the cropping system.

Pesticide Measurement

The ARMS provides cross-sectional data on pesticide use by producers who do and do not adopt GE crops for each year since 1996. Using the ARMS data, one could simply compare the mean pesticide use among current adopters and nonadopters of GE crops. However, differences in characteristics that affect the adoption decision may influence pesticide use decisions as well, making these simple comparisons suspect. The challenge is to find a way to control for all other sources of variation in pesticide use so that the adoption impact can be isolated.

Impact of Pesticide Use

The decline in pesticide use was estimated to be 19.1 million acre-treatments, 6.2 percent of total treatments. These estimates are associated with the changes in adoption that occurred between 1997 and 1998. While such changes would normally be small over a single year, the spectacular growth in biotech crop use meant that adoption increased by 160 percent for herbicide-tolerant soybeans, 150 percent for herbicide-tolerant cotton, 12 percent for Bt cotton, and 43 percent for herbicidetolerant corn.

Most of the decline in pesticide acretreatments was from less herbicide used on soybeans, accounting for more than 80 percent of the reduction (16 million acre-treatments). The estimated active ingredients applied to corn, soybean, and cotton fields also declined by about 2.5 million pounds, altough the total herbicide pounds used on soybeans actually increased as glyphosate was substituted for conventional herbicides. Once the results are broken down by type of herbicide, the data indicate that an estimated 13.4 million pounds of glyphosate substituted for 11.1 million pounds of other synthetic herbicides.

Toxicity and Persistence

The changing mix of pesticides that accompanies adoption complicates the analysis because toxicity and persistence in the environment vary across pesticides. The term herbicide or insecticide refers to a multitude of heterogeneous products. Thousands of formulations (commercial forms in which the pesticide is sold) are used. These formulations are mixtures of active chemicals (active ingredients) and inert materials, used to improve safety and facilitate storage, handling, or application. Hundreds of chemical products are used as active ingredients.

Each active ingredient has not only a different spectrum of pest control and potency, but also has a different impact on human health and the environment. Given this, it seems insufficient to report pesticide use by adding the quantities of all pesticides applied, even if expressed in pounds of active ingredient. For this reason, other measures are used such as acretreatments and proxies for the "impact" of pesticides on human health and the environment. Consider the most widely used GE crop, herbicidetolerant soybeans.

The adoption of herbicide-tolerant soybeans leads to the substitution of glyphosate for previously used herbicides. These crops are designed to allow farmers to limit herbicide treatments to as few as a single postemergence application of glyphosate, while a conventional weed control programme can involve multiple applications of several herbicides. In addition, and more importantly, herbicide-tolerant crops often allow farmers to use more benign herbicides. Glyphosate binds to the soil rapidly, preventing leaching, and is biodegraded by soil bacteria.

In fact, glyphosate has a half-life in the environment of 47 days, compared with 60-90 days for the herbicides it commonly replaces. In addition, glyphosate has extremely low toxicity to mammals, birds, and fish. The herbicides that glyphosate replaces are 3.4 to 16.8 times more toxic, according to a chronic risk indicator based on the EPA reference dose for humans. Thus, the substitution caused by the use of herbicide-tolerant soybeans results in glyphosate replacing other synthetic herbicides that are at least three times as toxic and that persist in the environment nearly twice as long.

Soil Losses

Availability, since the 1980s, of postemergent herbicides that could be applied over a crop during the growing season has facilitated the use of no-till farming practices, since weeds could be controlled after crop growth without tilling the soil. The use of herbicide-tolerant crops (particularly soybeans) has intensified that trend since it often allows a more effective and less costly weed control regime than using other postemergent herbicides. The impact of conservation tillage (including no-till, ridge-till, and mulch-till) in controlling soil erosion and soil degradation is well documented. By leaving substantial amounts of residue evenly distributed over the soil surface, conservation tillage

(1) reduces soil erosion by wind;

(2) reduces soil erosion by water (by reducing the kinetic impact of rainfall);

(3) increases water infiltration and moisture retention;

(4) reduces surface sediment and water runoff; and

(5) reduces chemical runoff.

Thus, by facilitating the use of conservation tillage (or no-till in particular), the adoption of herbicide-tolerant crops may reduce soil losses and runoff. However, there is little empirical evidence on how GE crops have affected tillage. Adoption of conservation tillage for soybeans grew (at a decreasing rate) from about 25 percent of the soybean acreage in 1990 to 48 percent in 1995, the 5-year period previous to the introduction of herbicide-tolerant soybeans. Growth of conservation tillage increased further in 1996, but then appears to have stagnated between 50 and 60 percent in the following years. A larger portion of the acreage planted with herbicidetolerant soybeans was under conservation tillage than was acreage growing conventional soybeans.

6

Potentials of Gene Revolution

The genetically modified crop movement employs a combination of previously unheard-of technologies in plant breeding, with the goal of improving agricultural yields. Unlike the Green Revolution techniques, Gene Revolution techniques involve cutting-edge biotechnological advances to achieve this goal. And for the most part, scientists in the developing world have not yet been trained in these technologies such that they can produce their own GM crop seeds. The term "biotechnology" encompasses all the techniques that use organisms, or parts of organisms, to produce or alter a product, or that develop microorganisms for specific purposes. By far the most prominent example of biotechnology in agriculture is genetically modified crops.

GM crops contain genes that are artificially inserted instead of the plants' acquiring them through sexual means. The transgene (artificially inserted gene) may come from a completely different species. Bt corn and Bt cotton, for example, contain genes from different subspecies of a soil bacterium Bacillus thuringiensis (Bt) that give plants the ability to manufacture their own pesticides. Depending on where and for what purpose the plant is grown, desirable genes may provide features such as higher yield or improved quality; pest or disease resistance (as with Bt crops); tolerance to heat, cold, or drought; or enhanced nutritional content-all desirable agricultural traits for farmers and consumers in the developing world.

Agricultural Benefits

Many genetically modified food crops are now planted around the world. The most common are GM soybeans, followed by GM corn, cotton, and canola. Others are GM tomatoes, potatoes, papayas, chicory, melons, rice, squash, sugar beets, and wheat. One of the most common and important purposes of GM crops today is to confer tolerance to herbicides that are sprayed on farmland to control weeds.

Herbicide-tolerant crops include transgenes providing tolerance to the herbicides glyphosate or glufosinate ammonium. These herbicides are broad spectrum, meaning that they kill nearly all kinds of plants except those that have the tolerance gene. Thus, a farmer can apply a single herbicide to his fields of herbicidetolerant crops, and can use the herbicide effectively at most crop growth stages as needed.

Another important benefit is that this class of herbicides breaks down quickly in the soil, eliminating residue carryover problems and reducing adverse environmental impacts. Another common purpose of genetic modification of crops is to confer protection against insect pests. Importantly, this crop trait could substantially improve yields in the developing world where pest damage is rampant and/or reduce use of chemical pesticides. The soil bacterium Bacillus thuringiensis in Bt corn and Bt cotton produces crystal proteins that are toxic to certain insects but generally harmless to vertebrates and non-lepidopteran insects. The genetic insertion of the bacterial gene into the plant genome enables the plant to produce its own pesticide.

Depending on the subspecies of the bacterium from which the gene is taken, the pesticide is toxic to insects of the orders Lepidoptera (which includes the common corn pests, European corn borer, Southwestern corn borer, and corn earworm), Diptera (mos- quitoes), or Coleoptera (beetles). Other purposes of modification in current and past GM crops include changing the fat and acid composition of the crop (canola and soybeans), conferring male sterility (chicory), conferring resistance to viral infection (papaya), and delaying time to ripeness to allow for longer transportation time and shelf life (tomatoes).

Future GM crops may be able to benefit farmers in other ways previously unrealised. Some GM crops may be able to grow in conditions that have been unsuited for agriculture. For example, scientists have developed and may soon be marketing a type of genetically modified tomato that is able to grow in salty soil. This type of modified tomato is particularly useful in areas where crops could not be grown previously because of soil salinity. The tomato itself channels the excess salt into its leaves so that the fruit retains its "normal" flavor. Other GM crops are now being developed that are well suited to drought conditions, that survive in soil heavy in metals such as aluminum, that convert or "fix" nitrogen from the air, and that produce vaccines against common diseases such as cholera and hepatitis B. Because farmers would thus be able to grow crops in otherwise difficult conditions, these crops, if affordable, could be an attractive option for farmers in areas with poor agricultural lands, notably in sub-Saharan Africa.

Most of the types of genetically modified crops grown today were produced with the intent to benefit agriculture. However, even these current types of GM crops have some indirect health benefits. For example, crops such as Bt corn that are genetically modified to produce insecticidal proteins have been shown in field studies to have lower levels of mycotoxins-chemical toxins and carcinogens produced by fungi. While the benefits of mycotoxin reduction in Bt crops are achieved primarily through reduced market losses in the industrialised world (i.e., farmers' crops will be more readily accepted if mycotoxin contamination is low), this benefit could have a significant health impact in the developing world, where exposure to food-borne mycotoxins is higher and few regulations exist to protect consumers (Wu, Miller, and Casman, forthcoming).

Moreover, because GM crops are more cost-effective for the farmers that adopt them, food prices go down, resulting in savings to consumers. Decreased food prices could then lead people to buy a greater variety of foods, thereby improving the nutritional content of their diets. Future GM crops could also have substantial direct nutritional or medicinal benefits to consumers. Crops could, for example, be genetically modified to produce micronutrients vital to the human diet. One type of crop in the making is "golden rice," genetically modified to produce beta-carotene, the precursor to vitamin A. Currently, about 400 million people in the world are at risk of vitamin A deficiency, which can lead to serious morbidity including night blindness, respiratory diseases, and even childhood death.

Rice modified with daffodil and bacterial genes would be able to produce beta carotene, which could in turn lead to a more balanced diet that could contribute to enhanced health in regions where rice is a staple of the diet. This type of crop would thus be potentially beneficial among Asian and African populations that suffer from malnutrition. Canola, too, can be genetically modified to enhance vitamin E content or to better balance fatty acids. Food crops engineered to produce edible vaccines against infectious diseases would make vaccination more readily available to children around the world.

Because of their palatability and adaptation to tropical and subtropical environments, bananas have received considerable research attention as a vehicle for vaccine delivery. Transgenic (i.e., GM) bananas containing inactivated viruses protecting against common developing-world diseases, such as cholera, hepatitis B, and diarrhea, have been produced and are currently undergoing evaluation. Because they would produce only the necessary antigens, these types of vaccine-producing GM crops may be safer than traditional vaccines whose additional materials often cause harmful side effects.

In the first decades of the Green Revolution, risks to human health and to the environment from the accompanying pesticides, fertilisers, and irrigation were not a significant concern. Consequently, the Green Revolution had little problem achieving the level of public acceptance that was necessary for it to have a revolutionary impact. But GM crops are a different story, due in part to the adverse environmental outcomes of the Green Revolution that eventually came under public scrutiny. The introduction of genetically modified crops into the food supply has generated a number of concerns about potential risks associated with this new agricultural technology. Depending on the seriousness of these risks, they may limit the GM movement's impact.

The risks fall under the broad categories of human health and the environment. When a new gene producing a novel protein is introduced into a crop, there is a chance that human subpopulations may have an allergic reaction to the protein. Also, GM crops may adversely affect non-target species; for example, a GM pest-protected plant targeting lepidopteran pests may spread toxin-containing tissues, such as pollen, that may contaminate the food of non-target lepidopterans.Cross-pollination between GM crops and non-GM crops, or GM crops and wild plant species, may occur, with unknown consequences.

Insects may eventually develop resistance to the insecticidal proteins produced by Bt crops, leading to inefficacy of both these crops and microbial Bt sprays. A host of other risks surrounding GM crops may exist of which we are currently unaware. Many of these risks, particularly those regarding food safety and impacts on non-target species, have undergone extensive scientific research. Most of this research has found no evidence that such risks exist among current GM crops. Indeed, 81 scientific studies financed by the European Commission have all shown no evidence of risk to human and animal health or to the environment from genetically modified crops. However, it is impossible at this stage to fully investigate all potential risks, current and future, of GM crops, and the challenges posed for such investigations by the ongoing development of GM varieties are substantial.

The Green Revolution has made it clear how important funding is to enable an agricultural movement to become revolutionary. In that case, financial support came largely from the public sector. Today, there is still a significant need to fund new agricultural research and technology transfer in the developing world. But biotech research and development (R&D) takes place almost exclusively in the developed world within private industry. Genetically modified crops are largely the product of private industry. This is partly because new technologies are far more costly than existing ones, and the biotechnology industry was able to gather the necessary funds to develop these technologies long before public awareness of GM crops could lead to publicly generated funding for GM crop development. Successful companies must focus on their markets with the intent of generating profit.

With regard to agricultural biotechnology, companies in the United States and elsewhere have thus far created primarily seeds that farmers in industrialised countries can and will purchase: corn and soybeans that can tolerate a particular herbicide, corn and cotton that are resistant to particular pests, and food crops that last longer on the supermarket shelf. Because of the "technology fee" that growers pay to use these crop seeds (including recoupment of industry's research and development costs as well as profit), and because the seeds are designed particularly for their planting situations, the targeted farmers in industrialised countries have generally found it worthwhile to buy these seeds and have been willing to pay the technology fee.

For this reason, there are currently two primary financial obstacles to using agricultural biotechnology in the developing world: (1) the lack of incentive on the part of the biotechnology industry to produce GM crops that are applicable in the developing world; and (2) intellectual property issues that discourage R&D by organisations, public and private, other than the firm first developing a technology. Because of these obstacles, GM crops that would be important to farmers in the developing world are in many places either unavailable or too expensive to be of practical use. Unless these crop seeds can be made more affordable, farmers will lack the incentive to use them and the chance of a revolutionary impact in the developing world will be small.

A fundamental challenge in this newest agricultural movement that did not arise during the Green Revolution is the definition and treatment of intellectual property (IP). IP issues are central to the Gene Revolution because whereas science and technology move forward through the sharing of ideas and resources, IP ambiguities and restrictions can often limit the valuable diffusion of science and technology. Commercial application of biotechnology has taken place primarily in the United States and primarily through the private sector.

The issue of who "owns" a particular event (the successful transformation) of a genetically modified crop and who can develop it further has become so economically important and contentious that numerous cases involving this issue are now being litigated. Some observers consider IP issues to be one of the most important impediments to the development and adoption of GM crops in the developing world. Several large multinational corporations that combine seed industries with chemical industries currently dominate agricultural biotechnology. These profit-driven companies have seen little reason to invest in expensive research and regulatory costs to produce crops that are grown on relatively few acres and must be heavily subsidised for poor farmers to afford.

Inevitably, private research focuses on needs of capital-intensive farming; research to feed the poor is less attractive, as it involves long lead times, risk related to unpredictable agricultural conditions and disregard of intellectual property rights, and beneficiaries with little capacity to pay. New agricultural technologies are usually expensive to develop because of the need for a large research infrastructure and the cost of meeting strict regulatory requirements. Therefore, corporations have focused their resources on creating biotech crops for farmers in industrialised nations, and not on biotech varieties of developing-world "orphan crops"such as cassava, millet, sorghum, cowpeas, or yams.

Because of the private sector's role in the financial organisation of the GM crop movement, IP considerations arise. Corporations own the rights to many of the necessary technologies and knowledge that lead to the development of useful GM crops; thus, public researchers in agricultural biotechnology as well as other private companies are often unable to access those technologies and knowledge that they need, or are legally blocked from using what they do know. Yet, it is important to remember that lack of intellectual property protection could make industry less willing to invest in developing countries. This is not to say that, left to its own devices, the private sector would ignore the concerns of the developing world altogether when developing agricultural biotechnologies.

In fact, some firms have recently expressed an interest in working with regional institutions to develop crops that would be beneficial to the developing world. In addition, they are willing to donate a substantial portion of their scientific knowledge, such as genomes of key food crops, for the purpose of increasing agricultural knowledge capital in the developing world. Moreover, several public-sector initiatives or partnerships between private and public sectors have emerged, with the technology transfer of GM crops to the developing world as one of their goals.

A notable example is the Rockefeller Foundation's African Agricultural Technology Foundation (AATF), which seeks to facilitate useful combinations of biotechnology with other methods of improving and empowering African agriculture. AATF is an African-led, Africanbased organisation designed to eliminate many of the barriers that have prevented smallholding farmers from gaining access to agricultural technologies. Genetically modified crops, as well as other tools of biotechnology, are one part of this goal; but importantly, AATF works toward an integrated approach to solving food security issues in Africa, which was previously left largely untouched by the benefits of the Green Revolution.

Another such example is the International Service for the Acquisition of Agri-biotech Applications (ISAAA). Supported by a combination of corporations, government agencies, and foundations, ISAAA is a nonprofit organisation that focuses on bringing GM crops to the developing world. However, studies on industrial decisionmaking in agriculture have shown that key determinants of the private sector's decision to invest in agricultural research are: (1) the perceived size of the market, and (2) the ability to reduce transaction costs when farms are larger. Most markets and farms in the developing world are small, and thus provide little incentive to the private sector to fund agricultural research that will help them.

Several important funding practices would need to occur for the GM crop movement to become revolutionary in the developing world. As the history of the Green Revolution makes clear, publicly funded research, if deliberately aimed at low-cost food production, can benefit developing-world farmers. Thus, one important component of the GM crop movement in the developing world is increased public-private-sector partnership. While this is happening to a small extent in the GM crop movement today, increased collaboration between these sectors would likely expedite the transfer of agricultural biotechnologies to poorer regions of the world.

Another important component of the GM crop movement is that intellectual property processes must be established to ensure appropriate protection and incentives to laboratories and manufacturers, while providing public researchers and developing-world farmers with access to the necessary tools for technology transfer. Such processes will allow GM crop seed to be affordable to the growers that would most benefit from them.

The Green Revolution had such a sizable impact because it took place in areas that were ripe for an agricultural revolution-namely, Asia and Latin America in the developing world and Great Britain among industrialised nations. Concerns about impending massive malnutrition, food security in the near future, and unwelcome political influences were all key pressures driving the need to adopt Green Revolution technology. Today, such needs are still present in much of the developing world, and most of all in sub-Saharan Africa. How has the GM crop movement been addressing these concerns? Thus far, the Gene Revolution has made its mark in only several places. Four nations-the United States, Canada, Argentina, and China-are together planting 99 percent of the world's total acreage of genetically modified crops.

China, in fact, was the first country to commercialise GM crops in the early 1990s, with the introduction of virus-resistant tobacco, followed soon by virus-resistant tomatoes. Some five million farmers have been growing GM cotton there for six years, with higher yields and without laborious and hazardous pesticide spraying. And as of March 2004, Britain has allowed the commercial planting of one type of GM herbicide-resistant corn. It is too early to forecast where around the world genetically modified crops will provide a benefit to local agriculture and where they might be adopted.

After all, GM crops have been commercialised for only about a decade, while Green Revolution technologies have been in use more than four decades. Even though there is a clear need for radical improvement in food production in certain regions, real obstacles stand in the way for agricultural biotechnology to meet that need. One obvious constraint is that the types of GM crops available today are not beneficial in many parts of the developing world. For example, the genetically modified herbicide-tolerant crops that are so popular in the United States and Canada are of little use in developing nations where farmers do not use herbicides on a regular basis.

Figure 1 shows that while the total acreage devoted to GM crops is steadily increasing in both industrialised and developing countries, GM crop acreage is much higher in industrialised areas. This is despite the considerably larger acreage of farmland in the developing world.

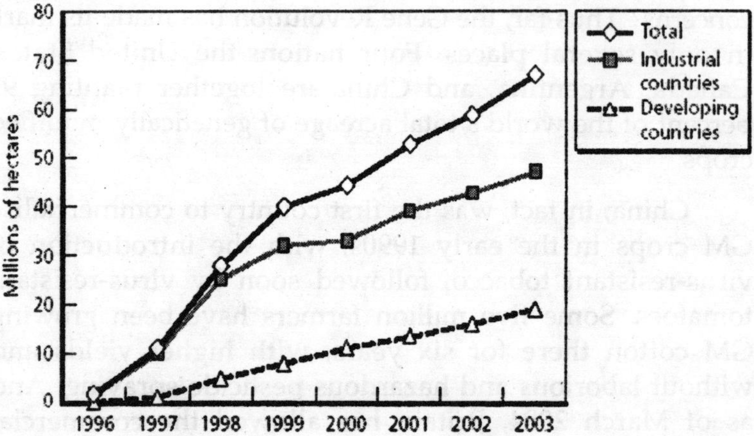

Figure 1. Area Devoted to Genetically Modified Crops, 1996 to 2003

Of those parts of the developing world where GM crops are grown, South Africa and India are the most prominent. In the Makhathini Flats of South Africa, for example, farmers have been growing GM cotton for four years. India, a nation that had previously shunned agricultural biotechnology, recently harvested its first Bt cotton crop (genetically modified to protect against insect pest damage). Qaim and Zilberman found a 79 percent average increase in Bt cotton varieties over conventionally planted varieties in India in 2002. If GM crop adoption in these nations is sustained over the next several years and possibly beyond, the GM movement may indeed be considered revolutionary in those areas.

Closely linked to the question of whether the GM crop movement becomes revolutionary in a particular region is whether the people of that region are willing to accept the new technologies and to adapt their lifestyles to meet the changes that the technologies may bring- criteria for a successful agricultural revolution. In terms of a people's acceptance of the new technologies, first, there are the direct impacts on food and culture. Compared with most Americans, people in many parts of the developing world place a higher level of importance on food as part of their culture.

Americans themselves felt wary about hybrid corn when it was first grown on U.S. farms in the 1920s and 1930s, so it is not difficult to imagine what the cultural impact might have been, or might be, in nations worldwide from the high-yield hybrid crop varieties of the Green Revolution, and even more so from genetically modified crops. To this end, GM crops have already become a stigmatised technology in many parts of the world because of a widespread fear of "Franken-food" and "playing God" with nature. Mistrust of industry may also play a role in these concerns. Second, GM crops may have indirect impacts that change the culture of a region more slowly, which will affect whether the GM crop movement can become sustainable in a revolutionary way.

Shifts in socioeconomic status due to increased food production and new agricultural methods can benefit certain groups in society but may cause other groups to suffer. For example, in the first third of the 20th century, machines increasingly replaced humans on farms, first in the industrialised world, then in parts of the developing world. But only those farmers who had access to sufficient land holdings to make the machinery useful had incentives to participate, and of those farmers, only those who could afford the purchase price could participate. Smaller farmers tended to suffer from the lower product prices resulting from increased production overall, and many sold or lost their land.

GM crops, besides being costly, may also favor those farmers with large land holdings-an outcome that would be unacceptable to many stakeholder groups. The lack of acceptability of this outcome on a regional basis, combined with the price of GM crops, could prevent the GM crop movement from achieving revolutionary impacts in some regions. Moreover, the new intellectual property issues associated with GM crops would require agricultural practices that differ substantially from traditional farming methods among subsistence farmers. Patent rights require farmers who want to plant GM crops to buy new seed every season. This requirement is not usually a problem in the industrialised world, because industrialised-world farmers have already been buying new seed for each season since their adoption of hybrid varieties.

Agribusiness in richer nations values crops that have uniform qualities (e.g., uniform height, weight, and development time), and these qualities are much more easily achieved through seed newly bought each year. But developing-world farmers are often not willing or able to buy new seed every season. About 80 percent of developing-world farmers currently save seed from the previous season in order to cut costs. Under current intellectual property stan-dards, a substantial change not just in practice, but also in the shift of cultural mindset from subsistence living to agribusiness, would need to take place for the Gene Revolution to succeed among developing world farmers.

Policies and Regulations for Genetically Modified Crops

Interestingly, the main constraint on the potential of GM crops in the developing world may not have to do with either agricultural conditions or credit availability, but with precautionary regulations. Policies and regulations for genetically modified crops have evolved very differently from those for Green Revolution technologies. In the years leading up to the Green Revolution, policymakers saw a political need for agricultural improvement worldwide, advanced this cause as a key policy concern, and ensured that funds were appropriated toward the cause. Plant scientists moved into foreign nations and aided regional farmers in adopting new agricultural technologies almost completely uninhibited by governmental regulations.

For the purposes of regulating the technology of genetically modified organisms (GMOs), a Coordinated Framework for Regulation of Biotechnology was created within the Office of Science and Technology Policy (OSTP) in 1986. However, its outcome was not so much to form new regulations for GMOs as to delegate responsibility for oversight to existing agencies under the framework of existing statutes. Importantly, the Coordinated Framework emphasised that risk assessment would be based on the biotech product itself, not the process by which the product was developed.

This regulatory approach, which has proven to be more permissive toward GM technologies with the direct impact of aiding the Gene Revolution in the United States, is quite different from the European Union approach, which emphasises the process. Hence, GM crops are regulated by three governmental agencies: the United States Department of Agriculture, the U.S. Food and Drug Administration (FDA), and the Environmental Protection Agency (EPA). Table 1 depicts the regulatory scheme for coordinating reviews of GM crops.

Usually, the first agency to evaluate the safety of new transgenic crops is the USDA. The Animal and Plant Health Inspection Service (APHIS) of the USDA has regulatory oversight for protecting U.S. agriculture from pests and diseases. There are two main steps to the GM crop approval process, one for field tests and one for commercialisation, under APHIS regulation.

Table 1. Regulatory Scheme for Coordinating Reviews of GM Crops

Product Class	Lead Agency	Federal Statutes
Plants	USDA/APHIS	FPPA, NEPA, PPA, FPQA
Pesticides	EPA	FIFRA, FFDCA, FPQA
Food and additives	FDA	FFDCA

First, registrants must obtain a permit to test the crop on controlled fields. APHIS issues permits after it has concluded that the GM plant in question exhibits no plant pathogenic properties; is no more likely to become a weed than its non-engineered parental varieties; is unlikely to increase the weediness potential for any other cultivated plant or native wild species with which the organism can inter-breed; does not cause physical damage to processed agricultural commodities; and is unlikely to harm other organisms, such as bees, that are beneficial to agriculture. Second, to commercialise a transgenic plant, the researcher petitions APHIS for nonregulated status-a status that has aided the Gene Revolution in the United States by its relief from regulatory restrictions from the USDA, except in the unusual case of proven harm.

While the U.S Food and Drug Administration (FDA) has broad regulatory authority to determine the safety of foods, food ingredients, and feed through the FFDCA, there is no particular statutory requirement regarding regulation of genetically modified foods. This is largely the result of the FDA's 1992 policy statement. In this policy, the FDA granted Generally Recognised As Safe (GRAS) status to GM foods, which meant that these foods did not need to undergo additional tests or analysis with the agency unless there was a claim of a nutritional or health benefit.

Thus, the FDA recommends, but does not require, that biotech industries producing GM crops undergo consultation with agency scientists before marketing their products. This ruling has greatly decreased the regulatory burden that GM crops undergo in the United States, again contributing to the ease with which the Gene Revolution has spread. Another action that has aided the Gene Revolution in the United States, but has caused controversy elsewhere in the world, was the FDA decision that GM foods do not require labeling.

The FDA currently requires labels on GM products only if the genetic modification introduces a protein from a food that is a known allergen (e.g., peanuts). However, because proteins produced by GMOs currently on the market are not known allergens, they need not be labeled under FDA policy. Furthermore, the FDA states that there is no basis to distinguish GM foods as a class separate from foods developed through traditional breeding.

By the authority of the Federal Insecticide, Fungicide and Rodenticide Act, the EPA regulates the sale, distribution, and use of pesticides, including those produced by GMOs. To be registered under FIFRA, a pesticide must not cause "unreasonable adverse effects on the environment." This statement has also facilitated the greater spread of the Gene Revolution in the United States as compared with the European Union because of the relatively lenient burden of proof. That is, GM crops are registered if preliminary tests show no evidence of unreasonable adverse effects.

To qualify for an exemption, the product must be shown to be nontoxic to mammals (including humans) and to have little potential for allergenicity in humans, based on in vitro digestibility tests, thermolability, and amino acid sequence comparison with an allergen database. In the final stage prior to full commercialisation of a product, the EPA reviews the application for registration on the basis of the following areas of required data:product characterisation, human health assessment, ecological assessment, insect resistance management, and benefits.

The European Union has adopted a very different regulatory approach from that of the United States to deal with GM food and feed products coming to market. In 1990, the European Economic Community (EEC) issued "Directive 90/220 on the Deliberate Release to the Environment of Genetically Modified Organisms," under the agency Directorate General (DG) XI for Environmental Protection. Three key reasons for the regulation were given in the preamble to Directive 90/220:

1. That precautionary action should be taken with regard to the environment,
2. That protection of human health and the environment requires that any risks from the release of GMOs to the environment should be controlled, and
3. That any variation between the member states concerning these rules would have an adverse effect on the operation of a single market.

In essence, the Precautionary Principle, which states in its most basic form that when an activity raises threats of harm to human health or the environment precautionary measures should be taken even if some cause-and-effect relationships are not established scientifically, became an explicit element of European regulation of GMOs, in contrast to the U.S. regulatory philosophy, which centers on evidence of unreasonable adverse effects. This Precautionary Principle has thus far halted the spread of the Gene Revolution in Europe, because of the inherent difficulty of proving that there is no risk involved with GM crops. Moreover, Directive 90/220 enables member states to adopt stronger measures if they wish to do so.

Some member states have created their own safety plans for GMOs, going far beyond the EU statutory basis.

For example, even after the EU approved Bt corn for registration, Luxembourg and Austria banned it, stating that Bt corn was an environmental and commercial threat to organic agriculture. When the European Commission (EC)attempted to remove these bans, only a few member states supported this attempt; therefore, the bans stayed in place. To this day, the GM crop movement has been blocked or delayed by individual countries despite EU approval; hence, there exists a de facto moratorium throughout Europe on the commercial growth, sale, and distribution of GM crops.

Part of the restrictions on GM crops in Europe stems from lawmakers' perception of strong public opinion against foods made from genetically modified materials and their perception that the public has a general lack of faith in the integrity of government regulators. This perception has been a limiting factor on biotechnological research and development. These anti-GM sentiments may be partly the result of other unrelated food crises that occurred in Europe in the mid-to-late 1990s, such as the outbreaks of "mad cow disease" (bovine spongiform encephalopathy) and more recently hoof-andmouth disease. Unfortunately for U.S. agriculture, the purported anti-GM sentiments in Europe have also led to a marked decrease in agricultural exports to the EU since 1996, when Bt corn was first commercially planted in the United States. These sentiments also indicate that it is unlikely that GM crops will prove to be revolutionary to agriculture in Europe, due to both the lack of acceptability among its citizens as well as lack of support from major governing bodies.

Recently, conflict between the United States and the EU regarding trade of GM crops has taken a new turn. With EU Directive 2001/18/EC,which came into effect on October 17, 2002, and a subsequent European Parliament ruling on July 7, 2003, the European Union ended its unofficial moratorium on GM crops by passing new legislation regarding the deliberate release of GMOs into the environment. Those rules came into force on April 18, 2004. Under the new rules, everything from breakfast cereals to animal feed with more than 0.9 percent genetically modified content will need to be properly labeled, which requires carefully tracing GM foods through the entire food system. Directive 2001/18/EC opens the way for biotech companies to apply for approval to market their products in EU countries, but American officials and others are doubtful that these rules will truly facilitate market access or encourage farmers to cultivate GM crops.

Tracing and labeling GM food requires grain segregation at the farmer and elevator level. Not only would farmers and grain elevator operators need to keep GM and non-GM crops separated, they would also have to prevent commingling of the two types of crops during harvest, transport, and storage, which would likely slow the rate of turnover in a high-volume business. Currently, elevator operators have very thin profit margins, and their profits depend on moving large volumes of product quickly. To use the example of genetically modified Bt corn, if all U.S. grain elevators were to adopt segregation of Bt/non-Bt corn, the total segregation costs across the United States would be expected to exceed $400 million annually. This amount is higher than the total estimated monetary benefits of producing Bt corn in the United States, even taking into account potential environmental benefits from decreased pesticide use and health benefits from mycotoxin reduction.

If these cropsegregation practices were to become the worldwide norm, even industrialised-world farmers soon may lose the incentive to plant GM crops because of high segregation costs, thus halting the spread of the Gene Revolution even in nations where farmers can afford the GM seed. While these costs are unappealing to American farmers and food processors, they are an even more serious problem in developing nations that lack the infrastructure to segregate GM from non-GM crops in the first place. This problem came to a head in 2002, when worldwide wariness of GM crops received a level of attention that the United States could not ignore.

For instance, although Zambia was facing a serious famine, Zambian officials decided to turn away 26,000 tons of U.S. food aid in October 2002, saying that the shipments contained genetically modified corn that was unsafe. A key reason for refusing the aid, according to Zambia's agriculture minister, was that the GM corn seed could pollute the country's seed stock and hurt its export markets. In fact, government officials in Zambia, Zimbabwe, Mozambique, and Malawi all feared that if some of the unmilled GM corn imported as food aid was instead planted by farmers their nations would lose their current status as "GM-free" countries, compromising their ability in the future to export any food and farm products to the European Union, or to receive aid from European donors.

Hence, the Gene Revolution has ground to a halt in these nations due to potential fears of anti-GM policies abroad. Zambia is not alone in its reluctance to allow any form of GM food-processed or unprocessed-through its borders. Surprisingly, China and India have also taken an anti-GM stance, despite past developments that had indicated their acceptance of GM crops. In 2002, India approved the planting of Bt cotton for the first time, and many expected that this movement would lead to a more widespread acceptance of GM crops in that country.

Two other important differences in the political worlds of the Green and Gene Revolutions involve nongovernmental organisations (NGOs), public awareness, and public opinion. Unlike NGOs at the time of the Green

Revolution, today's NGOs worldwide are an important political voice in the success or failure of agricultural revolutions. It was not until the 1970s and 1980s that NGOs representing environmental concerns gained full force. At least part of the motivation for the agri-environmental NGO movement stemmed from the drawbacks of the Green Revolution with its myriad pesticides and fertilisers that polluted local soil, water, and air.

Some of these environmental problems stemming from post-World War II agricultural advances were described in Rachel Carson's seminal work Silent Spring, first published in 1962. In the realm of the debate over the Gene Revolution in today's political climate, many NGOs have aligned themselves against worldwide adoption of GM crops. Since 1998, NGOs have been instrumental in forcing the EU to impose a moratorium on new GM crop approvals, and now they are working to prevent approvals in the developing world. Greenpeace in particular has invested $7 million to halt planting of GM crops, especially in developing nations that have not yet planted GM crops.

Furthermore, the public today is much more aware of what is going on in the agricultural world and food markets compared with the public during the days of the Green Revolution. It is not merely that genetically modified crops are more controversial, but also new information technologies (such as the World Wide Web) have greatly increased public access to coverage of such issues. Public surveys on the topic of genetically modified foods have helped to influence decisionmaking within the EU-in particular, surveys that revealed the desire on the part of citizens for the labeling of GM foods.

Indeed, the public information situation is dramatically different from the situation during the heyday of the Green Revolution, and it could prevent the Gene Revolution from having the same impact as the Green Revolution. The public opinion problem may be due to the fact that certain types of GM crops that were first introduced seemed to have no noticeable benefit to the general public. Taking Bt corn in the United States as an example, consumer surplus (i.e., the gain to consumers) through reduced market price of corn and corn products represents the largest total public welfare increase among stakeholders-about $530 million per year-a larger gain to consumers than to Bt corn growers as a whole. On a per-person basis, however, the consumer benefit amounts to less than $2 per year.

Although consumers are price responsive and genetically modified corn can result in overall lower average consumer prices for corn, the savings are not large enough to offset consumers' perceived risks and their rejection of GM crops. The political scene of the Gene Revolution has thus far proven to be dramatically different from that of the Green Revolution. The world is more complicated and stakeholders are more closely integrated. Food regulations in one part of the world could have a dramatic and often unintended impact on agricultural systems in another part of the world. The acceptance and even encouragement of new technologies that accompanied the Green Revolution are much more scarce in the Gene Revolution. Governments as well as consumers must accept GM crop technology if a Gene Revolution is to occur. Otherwise, there is serious doubt about how far the GM crop movement can spread worldwide.

7

Seed Production

Seed production is one of the least visible yet most important aspects of food and feed production. Seed production is a technical process that requires in-depth knowledge of the reproductive mechanisms in plants. Seed production has been defined broadly as the series of steps that begins with the breeding of a new variety and ends with seed that is sold to farmers.

Plant breeders develop varietiesand produce breeder seed-the source seed from which all varieties are propagated. In the case of corn, breeders will also be responsible for adding male sterility if needed. Parent or foundation seed production is usually done by a separate department. The production of hybrid or certified seed-the seed that will be sold to farmers-is done by yet another department.

Genetic purity is important because gardeners and farmers expect varieties with the same names to be identical from one year to the next. There are two sources of contamination: —impurity. resulting from physical mixing of seed and impurity resulting from the movement of pollen. The seed production process is described in sufficient detail to identify points where such contamination may occur. The goal is to determine if seed of pharma crops can be produced in a way that ensures virtually zero contamination of the food system.

Seed production practices for corn and soybean have evolved to provide farmers with genetically pure varieties. Four classes of seed are officially recognised in the United States: breeder, foundation, registered, and certified. A brief description of each, as taken from the AOSCA genetic and crop standards handbook, follows. Understanding these classes is important to understanding the U.S. seed production system.

1. Breeder seed is seed directly controlled by is seed directly controlled by the originator or sponsoring plant-breeding institution or person. This is the class of seed from which all other classes of seed are derived.
2. Foundation seed is developed from breeder is developed from breeder or foundation seed produced under control of the originator or sponsoring plant-breeding institution or person. Foundation seed is a class of certified seed produced under procedures established by the certifying agency.
3. Registered seed is developed from breeder or is developed from breeder or foundation seed and is produced and handled under procedures acceptable to the certify-ing agency.
4. Certified seed is developed from breeder, is developed from breeder, foundation, or registered seed and is produced and handled under procedures acceptable to the certifying agency.

Not all classes of seed are used in all crops, and it should be noted that the United States has no laws or regulations requiring that only certified seed be sold to farmers. Certification of seed production is optional for seed sold in this country but is usually required for seed sold in other parts of the world, especially Europe. The requirements for each class of certified seed vary from crop to crop. These are minimum standards and apply only to those segments of the flow chart in Figure 1.

Variety Types

The type of variety grown in a crop is determined primarily by the crop's mode of reproduction (selfpollinated in soybean versus cross-pollinated in corn). These are the major determinants of the steps involved in seed production. Seed production in corn is more complex than in soybean.

Of the many possible variety types, we will restrict ourselves to three: inbred lines, openpollinated varieties (OPVs), and single-cross hybrids. An inbred (or pure) line is a strain that breeds true to type when individual plants of the strain pollinate themselves (self-pollination). An inbred line can be developed from repeated generations of self-pollination or from repeated generations of crossing to a common parent (backcrossing). An OPV can be a landracedeveloped by farmers, landraces improved by breeders or farmers, or synthetics created by crossing landraces, synthetics, or inbred lines together.

OPVs are genetically heterogeneous and are maintained by allowing individual plants to cross-pollinate in isolation. A single-cross hybrid is the first generation cross between two inbred lines. OPVs were the dominant variety type in the United States before the introduction of hybrids in the mid-1930s. OPVs were gradually replaced by hybrids, and single-cross hybrids now account for at least 99 percent of the hybrid seed sold in this country. OPVs are no longer grown on large acreages, but when they are, seed production is usually done on the farm by the farmer.

The most popular variety type grown by farmers in soybean is a pure line. Pure lines of soybean are equivalent to inbred lines of corn. Soybean is predominantly self-pollinated, so inbred lines of soybean can be maintained by just planting and harvesting them. Seed mixtures (blends) of two or more soybean varieties are also a possible variety type. Hybrids have been proposed and have been evaluated extensively in soybean, but they are not widely used by farmers because of the difficulty and expense in producing hybrid seed.

Variety Development

Since transgenic varieties are not usually crossed together to initiate the breeding process, there is a virtually zero chance of a pharma crop contaminating the food system during the breeding phase as outlined in Figure 1. Transformation consists of three main phases:

— introducing the gene into plant cells;

— regeneration, the process by which a plant develops from the cells into which the gene was introduced; and

— maturation, the process of inducing the regenerated plants to produce seed.

Backcrossing is the breeding procedure used to move the transgene from the regenerated plants into other more desirable plants. The maturation phase of transformation and backcrossing is of primary concern for contamination.

Before the seed of a commercial single-cross hybrid can be produced, the inbred lines used to produce the hybrid must be developed. The methodology for developing inbred lines varies widely from breeder to breeder and from company to company. During the entire developmental phase, controlled pollinations are made either by hand or by wind in isolated crossing blocks similar to what is done in commercial seed production fields. The final products are inbred lines that can be used to produce a commercial hybrid.

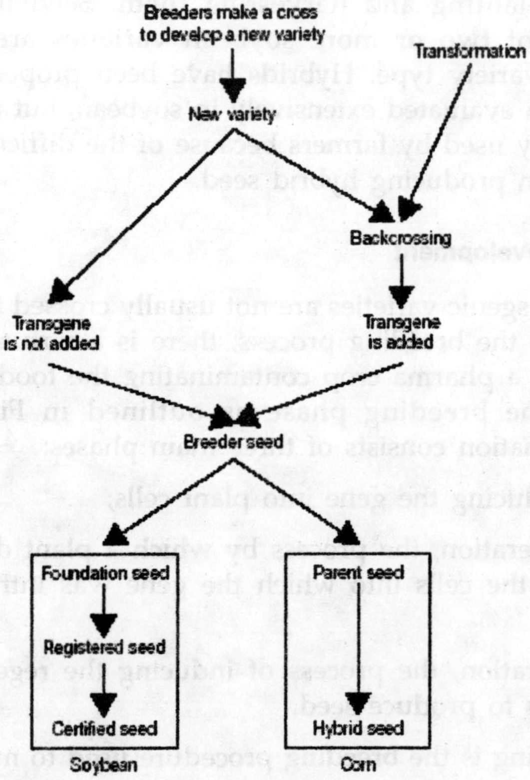

Figure 1. The Seed Production Process

The first step in generating a transgenic plant is to introduce a transgene into plant cells and have it become integrated into a cell's DNA. This is most commonly done in corn using a procedure called particle bombardment. Once cells contain the transgene, a plant is regenerated from the cells in cell culture. Regenerated plants are transplanted to a greenhouse so that when they flower, they can either be self-pollinated or crossed to another plant. The most difficult step in transformation is regenerating plants from cell culture. Because not all varieties can be used for this purpose, cell culture is said to be genotype-dependent.

Most corn transformation labs use the same basic genetic material for plant regeneration. This germplasm from which plants are regenerated is not commercially viable, so a backcrossing procedure is initiated immediately after plant regeneration to move transgenes of interest into commercially viable inbred lines. After plants containing transgenes (transformed plants) have been regenerated from tissue culture, molecular screens are conducted to identify individuals that carry only one or two copies of the transgene. Individual plant(s) meeting these criteria will usually be both self-pollinated and crossed to an elite line. These plants will be evaluated to determine if the transgene is inherited genetically and to verify that the transgene is producing the protein it was designed to produce.

Regenerated plants are self-pollinated, backcrossed to an elite inbred line, and crossed to a hybrid. The purpose of self-pollination is to obtain additional seed of the original plant that contains the transgene; the purpose of the backcross is to transfer the transgene into an elite inbred line. The subsequent cross to the hybrid enables the production of enough grain to begin preliminary protein extraction (in the case of pharma crops). All of these crosses occur in the greenhouse; all crosses in subsequent generations are made in the field.

The geographic area where these crosses are made is important. Breeding and backcrossing nurseries may be conducted on land rented or leased from farmers or on land owned by companies or universities. Some companies conduct their field breeding with experimental transgenes in dedicated off-season nurseries in other parts of the world. The use of such nurseries makes it easier to obtain isolation from other crops and minimises the probability of transgenes entering non-transgenic products outside the nursery.

Backcrossing is a breeding scheme designed to move single transgenes from one inbred line into another inbred line (Figure 2). The inbred line that the transgene will be transferred to is called the recipient. The recipient is typically an elite inbred already used in commercial hybrids. The inbred line or plant donating the transgene is called the donor.

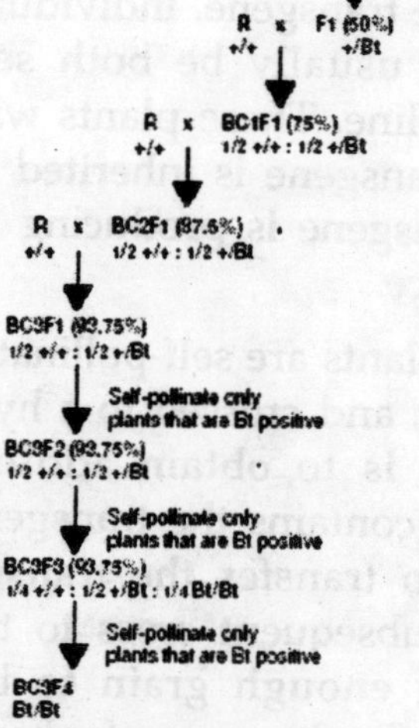

Figure 2. A Typical Backcrossing Scheme in Plants

The goal of backcrossing is to transfer the gene of interest from the donor into the recipient without otherwise genetically changing the recipient. Backcrossing starts by making a cross between the recipient and donor. The progeny produced from this and later generations are repeatedly crossed with the recipient inbred line.

With each generation of backcrossing, the amount of donor genome remaining is reduced by about one half. Each round is called a backcross. The number of backcrosses used varies from two to seven with an average of about five. Because the recipient inbred line is already used to produce commercial hybrids in corn, there is typically no field testing for performance until the desired number of backcrosses has been completed. The decision to conduct field tests for performance during the backcrossing procedure will depend on the amount of prior experience with the particular transgenic event.

After backcrossing is completed, the newly developed transgenic inbred line is crossed with another inbred to produce a hybrid for evaluation. The numbers of locations, replications per location, and years used for testing vary depending depending on prior knowledge of performance of the converted inbred, the quality of the conversion, and known problems with modifier genes associated with backcrossing a particular event.

Because a commercial hybrid receives extensive testing prior to commercialisation, it is not necessary to re-prove the performance of the hybrid. Following backcrossing of a transgene it is necessary to demonstrate that the backcrossing process itself did not disrupt the performance of the hybrid. Breeders are most interested in the comparison of the transgenic hybrid with its non-transgenic counterpart.

The amount of testing required to do this will depend on the protocols in place at specific companies and the amount of risk they are willing to take. The time lag between the development of a new inbred line and its sale to farmers as a transgenic hybrid is three to four years, depending on how long it takes to backcross the transgene, to obtain adequate quantities of parent seed, and to produce the hybrid.

Typical testing of a transgenic hybrid following backcrossing would involve 25 to 40 locations and one to three replications per location. The yield of pharma corn will not be such a major concern and less testing will need to be done. Comparison of a transgenic hybrid with its non-transgenic counterpart would continue until a decision was made to either drop the hybrid or use it commercially. If the hybrid will be used commercially, up to three years of data would likely be obtained on the hybrid before it is actually planted by farmers.

The backcrossing procedure is conducted by making hand pollinations. If the transgene of interest has been approved (that is, allowed on the market by the federal government), the necessary crosses are conducted in the field in summer or winter breeding nurseries used by the companies. Most companies have extensive protocols in place to keep approved transgenes from entering their non-transgenic breeding germplasm. Corn breeders at most breeding companies are required to use separate nurseries for GMO (genetically modified organism) and non-GMO breeding activities.

If the transgene is experimental and has not been approved, different procedures are used. Public sector scientists will conduct as much of the backcrossing as possible in the greenhouse, but it is also frequently done in the field using protocols developed by government regulators. Private sector scientists will usually do all of the backcrossing with experimental transgenes in a winter nursery. There are two advantages to this strategy: many winter nursery locations are tropical, enabling two or three breeding generations to be obtained in a year; and in many of these locations, such as Hawaii, it is easy to obtain long-distance (greater than one mile) isolation from other corn.Breeder seed of both transgenic and nontransgenic inbreds is established and maintained by the originating breeder or breeding company by hand pollinations.

Breeder seed of non-transgenic inbreds and inbreds that contain approved transgenes is maintained in summer or winter breeding nurseries as needed. It is the breeder's responsibility to ensure that the inbred is homozygousfor the desired transgene, has uniform plant type, and adequately represents the genetic constitution of the inbred. This description of corn variety development, transformation, and backcrossing is a generic description. There are numerous major and minor variations in the execution of all three activities.

Several breeding procedures are used to develop soybean varieties. Fehr lists five commonly used methods. The methods all involve self-pollination in every generation following the development of a breeding population. They differ primarily in the number of generations of self-pollination that are conducted before the lines are tested in replicated trials for their potential as new varieties. Single-seed descent is the most popular method of soybean variety development. Soybean is naturally selffertilising and hand pollinations are only used to produce hybrids for developing breeding populations.

The variety development phase for soybean, like corn, is not a source of contamination of the food supply by pharma crops. As in corn, the experimental transgenes and transgenes used in pharma crops are introduced into established varieties. Developing breeding populations that contain experimental or pharma crop transgenes would be too risky and difficult to control. The transformation phase for soybean is similar to that used in corn except that a different protocol may be used to introduce the transgene.If backcrossing is used to move the transgene into soybean inbred lines, the backcrossing procedure used would be nearly identical to that described above for corn (Figure 2).

Commercial soybean breeders use a procedure called forward breeding to incorporate transgenes (usually herbicide tolerance) into new varieties. This basically involves initiating new breeding crosses with inbred line(s) that already contain the transgene. If both inbred lines contain the transgene, all progeny resulting from the cross will carry it. If only one inbred carries the transgene, then only some progeny will carry it, and the breeders will have to select for the trait to keep it in plants that are self-pollinated. For herbicide resistance, selection is easy. The breeder sprays the soybean with the herbicide and keeps those that survive. There are three reasons why soybean breeders take this approach:

1) backcrossing is difficult in soybean because soybean is generally difficult to cross and backcrossing involves repeated rounds of hand crossing;

2) there is no time lag between the development of a superior variety and converting it to herbicide resistance; and

3) Roundup Ready soybean has been widely accepted and grown by farmers. No transgene has achieved such a wide acceptance in corn.

Since forward breeding is used primarily to incorporate herbicide tolerance into new soybean varieties, these breeding activities take place at standard soybean breeding facilities. Experimental transgenes are also often field tested in standard soybean production areas, judging by applications to the U.S. Department of Agriculture for field test releases. Many of these tests are conducted in Hawaii and Puerto Rico, but a number are also conducted in states where soybean is commonly grown. Breeder seed is produced and maintained in a manner similar to that in corn. The primary difference is that in soybean, controlled pollinations are not used to produce seed.

Breeder seed is produced by planting the soybean inbred line in the field and allowing it to self-pollinate.

Off-type plants, that is, plants that have different physical characteristics from the majority of the plants in the inbred line, are removed when they are identified.

Foundation or Parent Seed Production

Foundation seed increase is the next step in the process of developing a new variety (Figure 1). Foundation seed production is also called parent seed production in corn because the foundation seeds of inbred lines of corn are used as parents to produce the hybrid corn seed sold to farmers. Cytoplasmic male sterility (CMS) is a sterility mechanism frequently used in corn to reduce or eliminate the need to emasculate corn plants (remove the male part, or tassel) in the production of hybrid seed.

Large quantities of seed are required of each inbred line (the male and female parents of the hybrid) to produce enough seed for sale to farmers. This phase of seed production is usually referred to as foundation seed production. Foundation inbred seed production fields are planted with the initial supply of breeder seed. The foundation inbred seed increases are produced in isolation (via wind pollination) where the inbred line is allowed to sib-mateor open-pollinate.

Since all plants in the inbred line are theoretically identical, sib-mating or open pollination is equivalent to self-pollination. This initial inbred increase produces the first crop of foundation seed, which is then used to plant subsequent inbred increases and produce larger quantities of foundation seed for hybrid seed production. Foundation inbred seed increase fields are isolated from external pollen sources to maintain genetic purity. Only 1 in 1,000 off-type plants is allowed for certification of foundation inbred seed. Commercial companies certify most foundation seed that will be used for export or production of seed for export, but certification is not required for foundation seed produced for use in the United States. There is no registered or certified class of seed for inbred lines because this seed is not sold directly to producers.

Although larger companies produce their own foundation seed of inbred lines, smaller companies may contract this production to another company or simply purchase their foundation seed from one of the industry's genetic suppliers. Foundation inbred seed production has typically occurred in the primary corn production areas on a contract basis with farmers or on land leased from farmers. To increase purity, foundation seed production has recently been moving to areas with no commercial corn production.

If CMS is used to produce commercial hybrid seed, then parent seed production becomes slightly more complicated. The use of CMS involves an interaction between nuclear genes and genes inherited only through the cytoplasm. This system takes advantage of the fact that a plant can only inherit genes in the cytoplasm from the plant that served as its female parent. There are two types of cytoplasm in a CMS system, sterile (S) and fertile (F). There is also a dominant nuclear gene that restores the male fertility of plants that contain sterile cytoplasm. This dominant gene is usually called the restorer gene, and one copy is enough to restore complete male fertility. Inbred lines are converted to CMS using the same backcrossing procedure used to incorporate transgenes into inbred lines. In the case of CMS conversion, however, the objective is to substitute the whole genome of an inbred line into the sterile cytoplasm.

The production of parent seed from a malesterile inbred requires what is known as a maintainer line. A maintainer line is isogenicto the sterile inbred, carries non-restorer alleles, and has fertile cytoplasm. It is therefore male-fertile and will be used as the pollen parent in a seed production field similar to those used to produce hybrids, with the female being the sterile isogenic version of the same line. This is a slightly more complex process for producing parent seed than what is required if CMS is not used, but it must be done in isolation as well.

Production of foundation seed to be grown by farmers is much simpler in soybean than in corn. Once breeder seed is available, foundation seed production can begin. Foundation seed is produced in the same manner as breeder seed-by letting plants self-pollinate. The major difference between producing foundation seed in soybean and corn is that there is no isolation requirement in soybean beyond that required to prevent physical mixing. Although there is a registered class of soybean seed, it is not widely used and not recognised in some states. Soybean foundation seed is usually produced near the area where the commercial seed is expected to be sold.

Commercial Seed Production

Although corn and soybean have very different modes of reproduction, commercial seed production involves many of the same steps. The basic steps are seed packaging, planting, crop maintenance, controlling pollen movement (corn only), harvesting, transporting, drying, shelling (corn only), seed conditioning, bagging, and storing. Each of these steps often has several sub-steps and some of the processes differ slightly between corn and soybean. For example, there is no control for pollen movement in soybean, but isolation is used as insurance against physical mixing. Corn is harvested, transported, and dried on the ear, whereas soybean seed is removed from the pod by a combine in the field, then transported and dried. Corn seed production is technically more difficult and involves more steps.

Once the seed of a hybrid's inbred parents has been produced in adequate quantities, commercial production of the hybrid can begin. The first step is selecting the production area and the growers. The majority of hybrid seed corn sold here is produced in the Corn Belt region of the United States, but the seed of new hybrids is also produced in the Southern Hemisphere. One of the most important factors in selecting the growing area and grower is isolation of the seed fields from other corn fields.

Once the fields have been selected, the next step is planting. Production fields are planted using a predetermined ratio of male to female rows (Figure 3). Depending on the ability of the male to produce pollen and the difference in maturity between the female and male inbred lines, either one or two rows of male parent are used for every four to six rows of female.

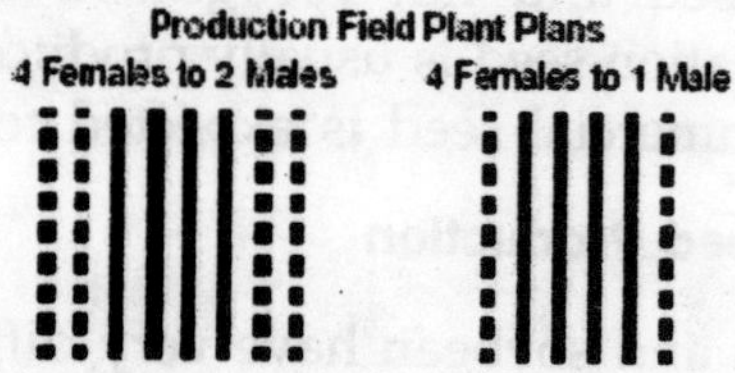

Figure 3. Typical Planting Arrangements Used in Hybrid Seed Corn Production Fields

The male and female parents may be planted at different times depending on the difference in their flowering times. In many production fields, once the male has finished shedding pollen it is mowed down to prevent accidental mixing at harvest and the theft of male inbred seed. Standard equipment and production and pest control practices are used at this stage.

Most cultural and management practices are designed to improve the yield of hybrid seed per acre. The primary concern in a seed corn production field is pollen control. Female parents of hybrids are prevented from shedding pollen either by emasculation or CMS. CMS may or may not require emasculation depending on the cytoplasm used, the genotype of the female inbred, and the environmental conditions during flowering. Neither emasculation nor CMS prevents contamination from pollen in nearby fields; this is accomplished entirely through spatial and sometimes temporal isolation. Ensuring an abundant supply of pollen from the male parent during the emergence of silks from the female also reduces contamination from nearby fields.

Once the grain in a seed production field has reached physiological maturity, the seed crop will be harvested on the ear to minimise damage to the kernels. Harvesting is done with special machines that minimise ear damage. The seed is then hauled to a seed production facility where is it dried, shelled, cleaned, sized, treated, and placed into bags. These operations are accompanied by a series of quality assurance operations to determine the physical quality, purity, and germination of the seed. The seed is then stored until its distribution to dealers and farmers. Certified seed is the class of soybean seed sold directly to farmers. It is either produced from registered seed or, more commonly, from foundation seed, and is usually produced near the area where it is expected to be planted.

Vulnerability of Seed Production

There are major differences in scale between seed production for commodity crops and pharma crops. On average, corn and soybean are each planted on more than 70 million U.S. acres annually. Predicting expected production acreage for pharma crops would be conjecture, but seed production needs will be far less than for commodity crops. This difference in scale implies that standard production practices may not necessarily be used for pharma crop seed production, which would therefore more closely resemble commodity foundation seed production. All seed production, however, no matter what the scale, shares some common steps and points of vulnerabilitythat will be explored below.

The situation is similar for soybean seed production, except that it generally takes more acres to produce enough seed to plant an acre of soybean. For example, planting 1,000 acres of soybeans would require 60,000 pounds of seed, which would take 25 acres to produce. Twenty-five acres of seed production would require about 0.6 acre of foundation seed production. In other words, producing enough seed to plant 1,000 acres requires about four times as much land for soybean as it does for corn. The frequency of seed production can also be a problem. For example, if only one pharma crop is to be produced on 1,000 acres, then seed production is greatly simplified.

If, however, 100 different pharma crops are to be planted on 1,000 acres each, then the frequency of production becomes a greater concern than the scale of production. The seed production system has been divided into six major steps (Figure 1): variety development, transformation, backcrossing, breeder seed production, parent seed production, and commercial seed production. Each of these steps is then analysed for points of vulnerability. Corn and soybean often share the same points, but the degree of vulnerability may differ.

The variety development stage of seed production is not a point of vulnerability because pharma crop transgenes are currently introduced only into established varieties. If, however, pharma crop transgenes were introduced at the variety development stage, and if the programme were conducted in parallel with conventional breeding programmes, then avoiding contamination entirely would become highly improbable if not impossible. Transformation is required to introduce all new transgenes into plants. The sources of potential contamination during bombardment and regeneration come from plant tissue.

Most transformation labs destroy all unused biological material by autoclaving, and the biological material is confined to a laboratory or greenhouse. As long as this process is followed, contamination of the food supply from biological material should not occur. The main concern with the transformation phase arises after a plant has been regenerated from cell culture and produces flowers. The first plants after regeneration are grown and pollinated in the greenhouse, where, without appropriate precautions, contamination of other corn plants with the new transgene via pollen movement or seed mixing is likely to occur.

The contaminated plants would most likely be other plants that already contain experimental transgenes. If the greenhouse is not a confined facility, pollen may escape into the atmosphere and contaminate plants outside the greenhouse. Detection of a contaminant by an outside third party at this stage would be difficult because the transgene and its product are usually regarded as confidential business information, which means that only the transgene developer could identify the contaminant gene. Detecting compounds produced by the contaminant gene is less likely to involve confidential business information, but the ease with which that can be done will vary. Backcrossing. The vulnerability of working with transgenes in the field is illustrated by backcrossing.

The primary use of backcrossing is to move a transgene into a more desirable inbred line, but all field-breeding activities, no matter their objective, have points at which food crops are vulnerable to contamination. The points of vulnerability are: seed packaging and preparation; planting the breeding nursery; making controlled pollinations; harvesting the breeding nursery; transporting, drying, shelling, and processing seed; field testing new varieties; and discarding the seed of unproductive varieties. Some of these activities occur simultaneously in a breeding programme. The main concerns for contamination during seed packaging are spilling and mixing, particularly if seed containing experimental transgenes is packaged in the same facilities with food/feedgrade seed. And there is always the risk of the planter not being properly cleaned before and after each use, especially if it is used for planting different types of corn.

The main problem during controlled pollinations, which are made either by hand or wind depending on the breeding procedure, is preventing pollen from moving to where it is not intended or wanted. Generally speaking, there are many more plants shedding pollen than are needed for use in backcrossing. The seed resulting from controlled pollinations in breeding nurseries is usually collected by hand and any seed left on the plants is harvested with machines. Harvesting involves several points of vulnerability. There are always plants that produce grain that is not harvested for seed.

The disposal of both this grain and the plants that produce the grain, as well as the cleanup of machines that glean the fields, poses special challenges. Seed may be spilled in the field during the harvest or during transportation to the shelling/threshing facility. The shelling/threshing process generates a lot of loose seed, as does the associated handling of the seed and spillage. Seed from different lots, for example, can be easily mixed or improperly labeled after shelling by inexperienced personnel. If the shelling/threshing machines are not properly cleaned, they too may become a source of contamination. One of the more serious problems during shelling is that there is frequently more seed than breeders need. This excess seed must be properly discarded so that it does not become mixed with other seed or grow the following year and produce pollen and more seeds.

Volunteer plants emerging in the field the following year are always a problem in breeding nurseries, where the frequency of dropped ears and partially shelled ears is high because of the variability among plant types. Field testing of new varieties poses many of the same vulnerabilities as described for variety development. The main difference is that field tests are more frequently conducted on land rented or leased from farmers. The field test may be located some distance from the breeding nursery, making daily or even weekly monitoring of the test site difficult. Otherwise, the vulnerabilities of field testing are almost identical to those encountered during the variety development phase.

Breeder seed is produced in breeding nurseries similar to those used for variety development. Therefore, the points of vulnerability are similar to those described above for backcrossing. The amount of breeder seed produced varies considerably depending on the circumstances, but a typical range would be 5 to 50 pounds per variety. Breeder seed is produced by making hand (controlled) pollinations.

Although the sources of contamination of breeder seed are the same as for backcrossing, the probability of detecting the contamination and its consequences are much different in breeder seed production using good practices (e.g., iterative self-pollination and observation). During the observation phase, mixing and outcrossing is nearly always detected before it can cause problems. Such a breeder seed production system is less vulnerable to contamination than backcrossing and other research operations.

Foundation seed production is the first large-scale seed increase that occurs in the seed production process.

The amount of foundation (parent) seed of a corn inbred line produced will depend on the anticipated sales of the hybrid for which that line is a parent. For example, if the hybrid is expected to be grown on about 10,000 acres, then about 0.4 acre of foundation production will be needed. In practice, the acreage may be larger if a multiple-year supply of seed is desired, if the inbred line is used as the parent of more than one hybrid, or if there are facilities in which to store the seed. The points of vulnerability for foundation seed production are similar to those described for breeding and field testing; the primary difference is scale and frequency.

Foundation seed increases in corn are almost always done in isolation and pollination occurs via wind movement. There is no easy way to contain corn pollen, and plants must produce pollen in order to produce seed. Because of the need to produce foundation seed for a large number of inbred lines, the steps following pollen movement become even more important. Harvesting, transporting, drying, shelling, conditioning, and storing would all be problematic if the same equipment and facilities were used to process seed.

If pharma corn were processed in the same manner and the same facility as commodity corn, the vulnerability to contamination due to seed mixing during these steps would be very high. The use of separate equipment and facilities is feasible for pharma corn. Commercial Seed Production. Hybrid seed production poses some of the same vulnerabilities to contamination of the food system as foundation seed production. The most obvious source of potential contamination is pollen movement from the production field to a farmer's field or a field producing another hybrid. Physical mixing.associated with harvesting equipment is another source.

Hybrid seed production also has the same contamination risks as on-farm production and risks similar to those for shipping and storage. Seed processing has several points of contamination, including the mixing of seed lots, imperfect cleaning of equipment in the processing plant, and mislabeling of bags. Although seed production plants are designed to maintain the purity of individual hybrids, virtually zero contamination is not necessary under normal circumstances and is never achieved. Separate equipment and facilities for pharma corn are feasible and could prevent cross-contamination with seed of varieties used for food..

Pharma soybean has far less risk of contaminating the commodity crop via windborne pollen than corn. In other ways, however, soybean can pose greater risks than corn. For example, soybean is visited by insects that facilitate pollination. Bees have the potential to travel long distances before depositing pollen, though the success of such long-distance pollination depends primarily on pollen longevity.

Otherwise, soybean has similar vulnerabilities to corn when it comes to contamination risks, including issues related to transformation, backcrossing, foundation and commercial seed production, and seed processing. Because a larger volume of seed is required to plant an acre of soybean than an acre of corn, there may be a greater risk of seed mixing with soybean than with corn. Commodity corn and soybean share many of the same points of vulnerability. The major points are pollen movement, physical mixing of seed, and seed left in the field to become volunteers the following year.

Pollen Movement

The fact that soybean is predominantly selfpollinated has led many to believe that contamination from outcrossing in soybean is not a concern,whereas the wind-pollinated nature of corn suggests that contamination from pollen movement cannot be eliminated. Surprisingly, some fundamental biological characteristics of corn make pollen control easier than in soybean production, particularly with regard to achieving the stringent goal of virtually zero contamination. Soybean produces between 3,740 and 7,600 pollen grains per flower.

If the average soybean plant produces 100 flowers, then that plant produces between 374,000 and 760,000 pollen grains. Since soybean is typically planted at the rate of about 150,000 plants per acre, the number of pollen grains produced per acre is around 75 billion. This is less than that produced by corn, but obviously a large amount nevertheless. Rates of outcrossing vary from less than one percent to more than 25 percent in soybean depending on the variety, environment, and availability of pollinators. Because little is known about insect-mediated long-distance pollen movement in soybean, spatial barriers may not be effective. Temporal barriers are ineffective because soybean tends to flower over long periods of time.

Mechanical and biological barriers to pollen movement are also unavailable in soybean, which is very difficult to emasculate manually and could never, therefore, be emasculated on a large scale. Biological sterility systems such as CMS only recently became available for soybean, and producing seed on male-sterile soybean plants is difficult. For these reasons, the production of pharma crop soybean seed in areas where commodity soybean is grown cannot be recommended until more is known about insect-mediated pollen movement. More tools are available to control pollen movement in corn than soybean.

Though spatial and temporal barriers are not completely effective, there are good mechanical and biological barriers available. For example, corn can easily be emasculated on a large scale either manually or with special machines; the key is to start before the beginning of pollen shed. Excellent CMS systems, which have been used for commercial seed production, are also available for corn. In theory, combining these four barriers-spatial, temporal, mechanical, and biological-could effectively achieve virtually zero contamination from pollen movement in corn. This assumes, of course, no human errors and a 100 percent effective CMS system.

The ability to use all four barriers effectively to achieve virtually zero contamination from pollen movement depends on the size of the fields, the amount of time devoted to monitoring fields for potential problems, and the isolation distances. It is difficult to quantify the probability of a contamination event. There are, for example, no published data on the frequency of failure for emasculation or CMS-this is an area that needs additional research on both the plant side and the modeling and simulation side. Strong qualitative statements are also difficult to make because they require too many assumptions about processes under human control and thus depend on measurement of human error rates.

Another possible approach to achieve virtually zero contamination is for farmers to plant inbred lines rather than hybrids. The male-sterile inbred could then be pollinated with normal corn, and commercial production could proceed in much the same way it does for hybrids. The main disadvantage of this method is that larger seed production acres of the pharma inbred would probably be required, since inbreds yield substantially less than hybrids. Growing inbred lines eliminates the hybrid seed production step but requires much larger increases of the inbred line carrying the transgene of interest.

Since there is no easy way to produce this quantity of inbred seed other than by allowing it to open-pollinate in isolation, inbred seed production would require greater isolation distances than CMS production and may need to be conducted in areas where the commercial crop or a seed production crop is not grown. CMS could still be used, but it would require the creation of the appropriate sterile and maintainer lines as described earlier. The type of hybrid corn seed produced for pharma crops is also important in preventing pollen movement. Assuming that the pharma transgenes are dominant, the hybrids to be grown by farmers might need to be 100 percent malesterile.

This can be accomplished in one of two ways. A male-sterile hybrid that is hemizygous for the transgene could be produced. The female inbred of the hybrid would be homozygous for the transgene and reside in sterile cytoplasm without nuclear restorers. The male inbred of the hybrid would be produced with a normal inbred as the male that lacks both the transgene and restorer alleles. The resulting hybrid would be male-sterile and hemizygous for the transgene. The farmer would plant the field in the same manner as a seed field with the male being a normal hybrid (Figure 3).

After pollination, the male rows in the field could be destroyed so they are not inadvertently harvested. The advantage of this system is that pollen would not be shed by transgenic plants.

Physical Mixing

Accomplishing this would require rigorous standard operating procedures (SOPs) to track the flow and quantity of seed and the use of dedicated machinery and infrastructure. The use of SOPs would ensure that employees were aware of the requirements, allow accidents or mistakes to be traced, and enable the process to be stopped or changed to prevent mistakes.

Volunteer Plants and Crop Residue

Volunteer plants are often a significant problem with regard to pharma crop contamination of the food system. If, however, dedicated seed production facilities and locations are used and no commodity crops are grown in that location, volunteer plants, even if they occur, would not have opportunities to contaminate commodity crops. Crop residues are also an issue in that they can be moved from field to field by equipment and wind.

8

Alternative Pharmacological Crops

Characteristics of Pharma Crops

The crop should have little use as a food or feed crop; an ideal pharma crop would have no role as either: Crops not used for food or feed would present few opportunities for contamination of the food and feed supply. Unfortunately, few crops are produced solely for non-food/feed purposes. All five of the world's major commodity crops, for example, are substantially or primarily used for human food. Wheat and rice are used mostly for food in all parts of the world.

Corn is used primarily as a food crop in the developing world, but in the United States, it is predominantly a feed crop with some important food (e.g., high-fructose corn syrup) uses. Soybean is primarily a source of oil (food) and protein (feed), but it also has other food uses. Cotton is grown primarily for industrial (fiber) purposes but also has feed (protein) and food (oil) uses. Thus, pharma crops will have to be found among those relatively rare crops used primarily for industrial purposes or among plants not yet adapted to agricultural use.

The crop should lack sexual reproductive organs or be amenable to processes that restrict pollen and seed dispersal: Even where pharma crops are not food crops, the dispersal of genes through pollen and seeds to other crops or wild plants is undesirable. The best pharma crop candidates would be plants that are entirely propagated vegetatively, or which produce pharma proteins in vegetative organs (e.g., leaves) that can be harvested before the appearance of any reproductive structures. Biological barriers to dispersal by pollen or seed would enhance the attractiveness of candidate plants. A self-pollinated reproductive system is preferable to a cross-pollinated system (though not sufficient). The availability of male sterility mechanisms, such as cytoplasmic male sterility, is also desirable, as is the absence of interfertile wild or weedy relatives.

The organ or tissue in which the pharmaceutical or industrial compound is produced should be easily stored and amenable to drug purification: The pharma compounds should be easily extractable without interference or contamination by other, natural metabolites in the organ or tissue in which the compound is produced. In addition, a plant tissue or organ that can be stored gives manufacturers flexibility. The availability of tissue-specific promoters to direct production in tissues that are not harvested for food or feed and not involved in reproduction (i.e., pollen, fruits, and seeds) would also be advantageous. For example, one could envision the production of pharma compounds in the leaves of potato plants but not in their tubers.

Appropriate production infrastructure should be available: The ready availability not only of the necessary equipment for planting and harvesting but also knowledge of the best production practices would shorten the time needed to develop new pharma crops.

The molecular information and tools needed to direct the production of pharma compounds at desired levels in the intended tissue should be available: This includes the availability of a transformation system to create transgenic plants, information about sequences necessary to direct tissue-specific expression, and knowledge of codon usage to ensure efficient expression by the cellular machinery. Maximizing gene expression can increase yields and minimize the land area necessary for pharma crop production. The following paragraphs discuss three possible alternative pharma crops in the context of these five characteristics.

Tobacco (Nicotiana tabacum Nicotiana tabacum, Solanaceae)

Tobacco is not used for food or feed, but is used in various products that humans inhale or inadvertently ingest.

Restriction of reproductive structures: The feasibility of leaf production reduces dispersal by seeds and pollen (i.e., plants can be harvested before flowering). Gene flow could be further reduced if expression of pharma transgenes were excluded from the seeds. Tobacco is a self-pollinating crop with no known sexually compatible wild relatives in the United States. It does not readily establish free-living populations, although it may sometimes-as in Hawaii-escape cultivation. Male sterility can be introduced via sexual crosses with wild Nicotiana species and has been genetically engineered via the nuclear genome. Ease of drug purification.

Pharma production in tobacco would likely take place in the leaves because the seeds are very small. This may or may not be a disadvantage, depending on whether the pharma compound can be stored in dried leaves. Although many tobacco varieties produce high levels of alkaloids (potential contaminants), there are low-alkaloid varieties that can be used for the synthesis of pharmaceutical proteins. Moreover, alkaloids are easily separable from proteins during purification.

Availability of production infrastructure: Tobacco is an excellent biomass producer (in excess of 40 tons of leaf fresh weight per acre based on multiple harvests per season) and a prolific seed producer (up to one million seeds per plant), thus hastening the time in which a product can be scaled up and brought to market. The existence of large-scale processing infrastructure is yet another advantage for tobacco.

Availability of molecular tools: Tobacco is easy to genetically engineer and is widely used as a model system to test the suitability of plant-based expression systems for producing therapeutic proteins and other transgene products. In fact, more transgenes have been introduced into the tobacco chloroplast or nuclear genomes than all other crop species combined. Fischer and Schillberg provide a recent list of pharma products produced in transgenic tobacco plants.

Both nuclear and chloroplast genomes have been transformed with very high efficiency, and several methods of genetic modification are readily available. Because chloroplast genomes are inherited predominantly through the female parent, the ability to splice pharma transgenes into chloroplasts limits the spread of the transgenes by pollen. The high levels of transgene expression via the chloroplast genome also minimizes acreage. For example, it has been recently shown that one acre of chloroplast transgenic tobacco plants could produce 400 million doses of anthrax vaccine.

Seed sterility has also been developed in tobacco. Overall, tobacco is the best prospect for an alternative pharma crop. Guayule (Parthenium argentatum Parthenium argentatum, Asteraceae, pictured on back cover) Use as a food crop. There are no known food or feed uses for this crop, which is the only crop other than the rubber tree that has been grown successfully on a commercial scale for the production of rubber. Like the rubber tree, latex is extracted from the bark of guayule stems.

Although commercial guayule operations have not had long-term success, the plant has recently received new attention because its latex may be less allergenic than latex from the rubber tree. Additional uses for guayule have also been recently identified; its fibers, for example, have been found to have insecticidal and fungicidal properties. The composite wood made from the residues remaining after latex extraction resists termite and woodrot damage.

Restriction of reproductive structures: This plant is a small perennial shrub adapted to the hot and dry conditions typical of the Chihuahua desert on both sides of the U.S./Mexico border. Although guayule exists as both diploids and polyploids within the species, tetraploids (2n=72) are the most common and usually the largest and most productive form. Guayule is partially apomictic and can have both sexually and asexually formed seeds on the same plant. The type of reproduction may be under environmental control. There are some 15 related species, several of which may hybridize with guayule. Latex content of the stem varies from 1 to 10 percent. Ease of drug purification. There is no information on the ease of purifying pharmaceuticals from the guayule plant.

Availability of production infrastructure: The agronomic characteristics of guayule need to be improved. For example, seed shattering at maturity is still a common occurrence. There are no commercial varieties available, but improved interspecific germplasm with increased aboveground biomass yield has been developed. Several agronomic aspects have been investigated, such as regrowth capability following harvest, oil and rubber production, and response to irrigation. Although production infrastructure is limited, there is interest in and experience with growing this crop.

Availability of molecular tools: Guayule can be transformed by Agrobacterium. Adoption of guayule as a pharma crop would probably require at least 5 to 10 years before full production.

Jojoba (Simmondsia chinensis, Simmondsiaceae, pictured on back cover)

The jojoba plant is a shrub reaching up to 15 feet in height, native to the Sonoran Desert, and adapted to dry and hot conditions. Individual plants have a lifespan of 100 to 200 years. Jojoba is an oilseed crop with up to 50 percent of its seed weight consisting of vegetable oil with unusual properties. Specifically, jojoba oil is a polyunsaturated liquid wax of a type not easily synthesized commercially. The only other source for this kind of oil is the sperm whale, which was once killed in great numbers for its oil.

A ban on the importation of sperm whale oil led to the discovery that jojoba oil is in many ways superior-it has high viscosity, a high flash and fire point, high stability, low volatility, and is non-toxic, biodegradable, and resistant to rancidity. Use as a food crop. Jojoba has a few food uses; it is primarily an industrial and feed crop. Its oil has been formulated into lubricants, cosmetics, pharmaceuticals, waxes, animal feed supplement (20 to 30 percent protein content of oil-less meal), and animal browse food. A small percentage of commercial production is used for salad oil, vegetable oil, and shortening. American Indians have long eaten soft-skinned jojoba nuts; the roasted nuts smell and taste like roasted coffee beans.

Restriction of reproductive structures: Pharma compounds would most likely be produced in seeds, although pollarded, aboveground biomass is a possibility. Confinement may be problematic for this species because it is cross-pollinated. Individuals are either female or male, the pollen is distributed by wind, and only 8 to 10 percent of the plants need to be male for sufficient pollination. On the other hand, natural distribution is restricted to parts of southern California, southern Arizona, and northwestern Mexico. Ease of drug purification. There is no information on the ease of purifying pharmaceuticals from the jojoba plant.

Availability of production infrastructure: Selection of jojoba clones with high yields and early seed production has been performed in several countries. Frost-resistant clones have been selected in Australia and the United States. Clones with different chilling requirements have been selected in Israel. Since seedlings are heterozygotes, the genetic variability is vast, and selection for desirable agronomic traits can be done in seeded plantations if necessary. Increase of desirable genotypes by vegetative propagation is easy. However, production infrastructure may be limited.

Availability of molecular tools: There is limited molecular information available for jojoba. Overall adoption of this crop for pharma production would require more than 10 years of research and development.

9

Traditional Seed Supply

Transgenic seeds or the transgenic seed supply refers to seeds used to grow transgenic crop varieties. The DNA sequences introduced into plants during the genetic engineering process are referred to as transgenically derived or transgenic sequences, and novel genes transferred to crops using genetic engineering techniques are referred to as transgenes.

Genetic Engineering

Genetic engineering has been a controversial technology from the beginning, especially in Europe and other countries outside the United States. Concerns about the use of the technology in agriculture have focused on a tangle of issues ranging from concerns about food and feed safety to environmental risk and corporate control of the food system. In theory, genetic engineering can modify plants to produce a wide range of new traits. Yet most engineered varieties commercially planted in the United States and around the world have been modified to express only two narrow categories of traits: resistance to a particular herbicide (thus permitting the use of that herbicide) or the expression of a pesticidal toxin derived from the soil bacterium Bacillus thuringiensis (Bt). These are (Bt). These are referred to as herbicide-resistant and insectresistant (or Bt) varieties, respectively.

Bt and herbicide-resistant versions of major crops were first planted on a large scale in 1996 and have been widely adopted in the United States during the last few years. In 2002, for example, about three-fourths of U.S. soybean acres, onethird of U.S. corn acres, and nearly 70 percent of North Dakota's canola acres were planted with engineered varieties. (North Dakota accounts for 89 percent of U.S. canola production.) Traditional, or non-engineered, crop varieties nevertheless remain popular as well, and U.S. farmers continue to plant them in large quantities. In addition to the handful of transgenes present in commercial varieties of herbicide-resistant and Bt crops, hundreds of other transgenes have been engineered into crops. These varieties, though not yet commercialised, have been field tested in the open environment.

Seed Contamination

Seed contamination is only one source of the contamination that bedevils exporters of nonengineered bulk grain and oilseeds. Most contamination is attributable to events that occur after the engineered and non-engineered varieties of seed are planted (Figure 1). There are two types of mixing events that occur after planting: physical mixing, such as commingling in grain elevators; and outcrossing, the movement of genes via pollen into neighbouring fields of sexually compatible crops. Since both of these phenomena are difficult to control under the current systems of grain and oilseed production, transport, and storage, mixing would occur even if the seeds farmers planted were absolutely free of transgenically derived sequences. While starting with seed contaminated with transgenically derived sequences exacerbates these problems, pure seed would not alleviate them.

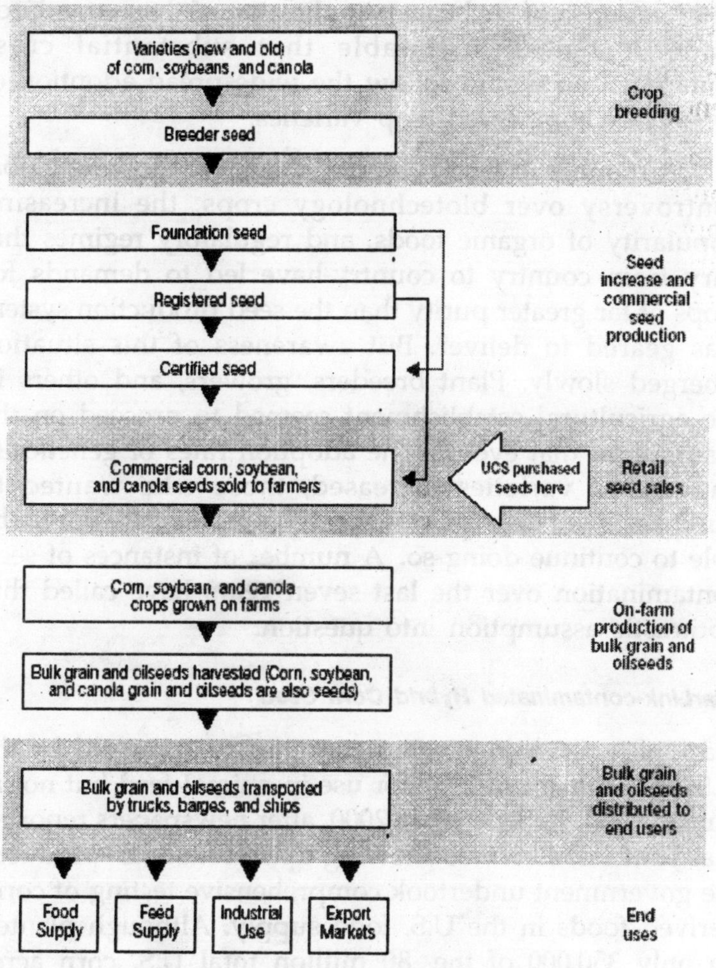

Figure 1. Seeds in Commodity Agriculture

Breeders working with genetically engineered varieties continued to use the same seed purity standards that applied to traditional varieties. Those standards vary from crop to crop but allow, in the case of soybeans, for example, up to 0.6 percent of the seeds to come from other kinds of crops such as corn and up to 0.5 percent from other varieties of soybeans. Application of these standards made it almost inevitable that substantial cross contamination would follow the widespread adoption of genetically engineered crop varieties.

A number of factors-among them, the growing global controversy over biotechnology crops, the increasing popularity of organic foods, and regulatory regimes that vary from country to country-have led to demands for crops of far greater purity than the seed production system was geared to deliver. But awareness of this situation emerged slowly. Plant breeders, growers, and others in the agricultural establishment seemed to proceed on the assumption that even as the adoption rates of genetically engineered varieties increased, those who wanted to purchase seed free of transgenic components would be able to continue doing so. A number of instances of seed contamination over the last seven years have called that optimistic assumption into question.

StarLink-contaminated Hybrid Corn Seed

StarLink was an engineered corn variety approved by the U.S. government in 1997 for use in animal feed but not in human food. In September 2000, after newspapers reported that StarLink corn was showing up in consumer products, the government undertook comprehensive testing of corn-derived foods in the U.S. food supply. Although planted on only 350,000 of the 80 million total U.S. corn acres (about 0.4 percent) in its most popular year, genetic sequences from StarLink corn varieties were eventually detected in numerous consumer products distributed throughout the U.S. food supply and in exported corn.

By 2001, StarLink also contaminated the U.S. corn seed supply. Fearing recurrent introduction of the illegal contaminant into food via the seed supply, the USDA instituted a programme to buy up corn seed that tested positive for StarLink. In June 2001, the department announced that it had already purchased $13 million worth of StarLink contaminated seed from 63 companies and was considering additional expenditures of up to $5 million. Despite concerted effort, it has proved surprisingly difficult to purge the U.S. grain system of the contaminant. As recently as December 2003, StarLink was still being reported in domestic grain. Part of the explanation may be that the seed supply for corn is still contaminated. It may be that inbred lines remain contaminated with StarLink genetic sequences and every time these inbreeds are used to produce hybrid corn seed, the StarLink sequences are reintroduced into the seed supply.

Contaminated Canola Seed

In 1997, Monsanto, a leading biotechnology company, recalled 60,000 bags of seeds of one of its transgenic canola varieties in Canada because they were contaminated with seeds of another transgenic canola variety (RT-200), which had not been approved for marketing in that country. Four years later, Monsanto detected the RT-200 contaminant again in seeds of commercial transgenic canola varieties in Canada. Even though RT-200 varieties had gained approval in Canada by that time, Monsanto withdrew the contaminated seeds from the market because the contaminating varieties had not been approved in all countries to which Canadian canola would be exported.

Monsanto admitted in 2002 that RT-200 seeds might also have been contaminating U.S. canola seed supplies since 1999. Even though the company has no plans to commercialise RT-200 in the United States, it sought approval of the variety in this country that year to minimise the disruption caused by its contamination of other canola varieties. In the spring of 2000, Advanta Seeds UK acknowledged that traditional canola varieties contaminated with an engineered variety (GT-73) had been sold to several European Union (EU) countries-where it had not been approved for sale-in 1999 and 2000. In 2002, Scottish scientists discovered that transgenic canola plants being tested in field trials were contaminated with a transgene not approved for testing in the United Kingdom.

Organic Food and Fiber

Organic food and fiber is one of the fastest growing sectors in U.S. agriculture. Not only do many consumers expect organic food to be free genetically engineered material, but federal standards also forbid the use of genetically engineered varieties in the production of organic foods. Organic growers seeking to meet this standard are finding it increasingly difficult to obtain non-engineered seed. Almost half the organic growers surveyed recently felt that contaminated seeds (rather than post-planting pollen drift, for example) represented the greatest source of contamination from engineered varieties. The difficulty in producing pure seed has led some organic seed companies to move their seed operations outside the United States.

While any one of these incidents might reflect an isolated example of seed contamination, taken together they reasonably suggest a more widespread phenomenon. The prospect of broad contamination of the seed supply raises important questions for food safety, international trade, organic agriculture, and the integrity of the seed system at the base of our global food supply. The growing evidence of seed contamination should have prompted the U.S. government to determine the extent to which seeds marketed as non-engineered are currently contaminated with engineered sequences.

Indeed, the Union of Concerned Scientists (UCS) and others in the public interest community have suggested the government undertake such a study. But it has not responded. So, UCS decided to conduct a pilot study of its own to assess the extent of contamination in the U.S. traditional seed supply. These seeds, along with seeds for transgenic varieties, are available from seed retailers, by mail order, and over the Internet.

Identity-Preservation Systems

The purity of seed is an issue of growing interest outside the arena of genetic engineering. New efforts are under way to create value-added markets for high-value crops, including some produced by genetic engineering. High-value crops exhibit desirable traits such as increased levels of important nutrients or the ability to produce a drug or industrial chemical. Today's commodity system, which minimises transportation, cleaning, and handling costs in part by tolerating a relatively high degree of cross-contamination, cannot meet the need for segregated, pure supplies of these high-value crops.

Spurred by market demand, individuals and companies are taking on the challenge of developing new infrastructure and delivery systems for value-added products. New "identity-preserved" systems create alternative pathways between seed suppliers, growers, and customers that avoid the current commodity system and its endemic sources of cross-contamination. The U.S. government is currently exploring ways to facilitate the marketing of identity preserved products. For example, the USDA is considering ways to reconfigure the commodity grain system to make segregation more feasible. Fundamental to the new systems devised to preserve identity" is the ability to produce and preserve the purity of seed.

The commercial seed stocks of non-engineered (traditional) commodity corn, soybean, and canola varieties are pervasively contaminated with low levels of sequences originating in genetically engineered varieties. The genes and genetic sequences we detected came from popular transgenic varieties currently allowed on the market in the United States. Until we know otherwise, it is minimally prudent to assume that any transgenes or transgenically transgenes or transgenically derived sequences being produced and field tested in the United States could move into the seed supply of corn, soybeans, canola, or any other crops with engineered varieties. The vulnerability of the seed production system to contamination is due primarily to its design and standard operating procedures. Contamination is likely to continue unless that system is changed.

Pharmaceutical and Industrial Crops

Agricultural biotechnology is entering a new age. No longer are researchers concentrating only on inserting genes that result in plants with traits like herbicide and insect resistance that make crops cheaper or easier for farmers to grow. Now they are inserting genes to create plants that produce drugs and industrial chemicals-in essence turning the crops into biological factories. The developers of the new pharmaceutical-producing "pharm" crops especially promise compelling benefits: new drugs that would otherwise be unavailable, and decreased production costs leading to lower consumer drug prices.

A wide variety of genes has been engineered into plants for pharmaceutical and industrial purposes. Many crops containing these genes have been tested in the open environment. Corn is the crop most widely tested for use as a pharm crop, but other food and feed crops including rice, potatoes, soybeans, tomatoes, and canola are also being used. Many other pharm crop transgenes have been tested but their identities are withheld from the public as confidential business information. The production of drugs and industrial chemicals in corn and other food crops presents obvious risks.

If genes find their way from pharm crops to ordinary corn, they or their products could wind up in drug-laced corn flakes. In addition, crops that unintentionally contain drugs or plastics could also prove harmful to domestic animals that eat contaminated feed; to deer, mice, birds, and other wildlife that feed in pharm crop fields; or to organisms living in the soil. The prospect of pharmaceutical genes contaminating the seeds we depend on for our food supply is genuinely troubling. If seeds are contaminated with genes for drugs, farmers will unknowingly plant and harvest what could be very dangerous crops.

The fact that many pharm crops will be planted on small acreage does not assuage the concern. The likelihood that seeds would become contaminated with genes from pharm crops is difficult to assess. It will depend on how the seed contamination occurs and a number of other factors, such as whether fields intended for seed production or seed increase for food and feed crops are located close to areas where pharm crops are grown. More study is needed to understand how often seeds are contaminated and where in the seed production process contamination occurs.

Pharm and industrial crops, for the most part, remain in the early phases of development. At this point, we should still be able to control the risks of this technology by imposing a strong new regulatory system. Now that we recognise that seeds could become contaminated with pharm or industrial products during the field testing phase and that these genes could make their way into commercial agricultural production, we need to ensure that the seed supply for food crops is explicitly protected in the development of such regulations.

Food Safety

There is no reason to believe that genetic sequences

originating in transgenic crops per se render food unhealthful. Only if the genes or their products cause problems on ingestion is there a food safety hazard, a determination that needs to be made on a case-by-case basis. A new, stronger system would inspire a higher degree of public confidence in the safety of engineered foods, particularly those products that will be brought to market in the future.

The system should be based on more rigorous science and include more tests for unexpected effects, as recommended recently by the Codex Alimentarius Commission, an international body that sets food safety standards under the auspices of the Food and Agriculture Organisation and the World Health Organisation. The U.S. National Academy of Sciences recently conducted a study on the hazards and unintended impact of engineered food on human health, and is preparing a report expected to contain recommendations for improving the food safety assessment process.

While most novel gene products will probably prove safe to consume as food and feed, such products are not inherently safe. For gene products that turn out to be harmful, the general concern in the seed contamination context is that the products will make their way into non-engineered seed varieties and be perpetuated in those crops by successive breeding cycles. The new products might not be readily identified as harmful because they would occur sporadically in products not recognised as genetically engineered. The degree of concern about potential contaminants of food and feed crops varies with their regulatory status and intended use.

Transgenic products that have undergone government scrutiny for use in or as food and feed (e.g., for herbicide and insect resistance) tend to raise the least concern. These products were developed for human and animal consumption and have at least been evaluated at some level and screened for obvious problems. Transgenic products that have not undergone food safety review but have been and are still being field tested raise more concern. Under the U.S. regulatory system, agencies do not analyse genetically modified crops for food safety until after they have undergone years of field testing.

This means that transgenic crops are potentially available to contaminate the seed supply long before a decision has been made about their safety. Examples of engineered crops that have been field tested but not evaluated for food safety include rice resistant to fungal diseases and corn with modified oils, starches, and proteins. Although not necessarily harmful, transgenic crop varieties that have not been scrutinised are of greater concern than scrutinised products because they have undergone no screen to remove dangerous transgenes.

Finally, gene products that are not intended for use in food raise the highest level of concern. They are unlikely to be reviewed for food safety at all, and many, such as pharm and industrial crops, are likely to produce bioactive and toxic compounds. Ad hoc accumulation of several novel genes raises food safety concerns.

In seed production systems that allow new genes to move into seeds via cross-pollination, every season offers new opportunities for the introduction of new traits. Single plants could accumulate and propagate several different novel traits over time, especially if they offer selective advantages. For example, in the short time that herbicide-resistant canola has been grown in Canada, genes for resistance to three different herbicides have accumulated in individual canola plants-whose offspring show up as weeds in fields planted with canola and other crops.

Two of the resistance traits originated in engineered canola varieties and one came from a traditionally bred variety. Whatever food safety dangers may accompany the presence of single novel genes, combinations of genes raise new concerns. The combinations of traits would not likely have been reviewed by agencies for food safety and may present synergistic or otherwise unpredictable effects. Accumulation (or the natural stacking) of traits is most likely to occur in crops whose seeds are routinely saved and planted. Parental lines of hybrid crops or true-breeding crops such as canola or soybeans fit in this category.

The additional risk posed by a transgene contaminating traditional varieties of a crop is likely to be small where the transgene is already present in widely planted commercial varieties of the same crop. Seed contamination, however, offers new routes by which transgenes might make their way surreptitiously to new environments-with unknown effects. Just as with food safety, the presence of engineered traits in the supply of traditional seeds is not necessarily a problem from an environmental perspective. Nothing about genetic engineering suggests that in and of itself, gene products derived from transgenic crops constitute an environmental threat or that engineered sequences inevitably render non-engineered plants dangerous to the environment.

But, again, neither are all such crops inherently safe; some do present environmental risks. The nature and degree of these risks depend on the traits added, the plants to which they are added, and the environment within which the plants are situated. Environmental risks are complex in nature and highly context-dependent. The main environmental risk of genetically engineered crops is that they would become weeds or transfer traits to wild relatives that would become weeds. Whether crops become or give rise to weeds depends on the genes they carry and, importantly, where they are grown. Crops cannot contribute genes to wild and weedy relatives if none exist nearby.

One question here is what additional risk to the environment is posed by a transgene present as a contaminant in traditional varieties of a crop beyond the risks posed by the growth of commercial varieties containing the transgene already permitted in commerce. In general, as long as the level of contamination remains low, where the transgenes at issue have been allowed on the market and the varieties containing them are widely adopted, the increased exposure due to the contaminants in the seed supply is unlikely to substantially increase exposure to the transgenes or the overall risk. The increase in the levels of Bt toxin coming from contaminated corn seed, for example, will not add much to the overall pressure on the environment resulting from the stream of Bt toxins already in the environment due to commercial Bt products.

On the other hand, seed contamination offers genes and gene products surreptitious paths to new environments. In most cases, neither seed sellers nor farmers would be aware of the contaminant, which would undermine their ability to effectively manage for environmental risks. The greatest risks would be associated with untested or disallowed genes, but even allowed genes might be a problem. For example, transgenic salt-tolerant rice might be commercialised under conditions designed to keep the plants from invading coastal wetlands. If the transgenes for salt tolerance were to contaminate the traditional rice seed supply, however, their presence would not be known and no precautions taken.

The contaminated rice seeds, finding their way to wetlands, could upset a delicate and important ecosystem. Furthermore, if the traits conferred a selective advantage, this would increase the prevalence of the transgenes in the new system. It is also possible that plants from the transgenic contaminating seeds could breed with wild plants, transferring new traits into wild populations. The effects of such transfers depend on the traits, the receiving populations, and environmental pressures and stresses. But harmful effects are certainly possible.

The possibility that some of the large number of transgenes that have already been field tested in more than 40 food and feed crops may already be moving into wild plant populations is troubling. In contrast to food and feed safety concerns, the relatively low level of contaminating transgenes found in any particular seed batch is not a limit on the amount of harm these transgenes can do in the environment. Considerations of ecological risk must take into account the ability of favourable environments to select for and increase the proportion of harmful transgenes in plant populations.

Contamination with transgenes from pharm and industrial crops raises environmental issues of special concern. These genes may be the sources of toxins that harm wildlife. In addition, toxin production is a common strategy by which plants protect themselves from predators, and pharm genes may provide selectable advantages in wild plant populations. If such transgenes are transferred from pharm crops to weedy relatives or used in crops that have tendencies to become weeds, they may enable crops to become weeds or make existing weeds more resilient and difficult to control. For example, aprotinin, a cow protein that has human medical uses but is also an insect toxin, has been produced in engineered corn plants.

If aprotinin genes were to move from aprotininproducing pharm crops into weedy relatives, the new genes might make the weeds hardier by enhancing their ability to withstand insect predation. The likelihood of pharm genes establishing themselves in weedy populations is enhanced where the pharm genes confer an advantageous trait such as insect resistance. Seeds contaminated with Bt insect-resistance transgenes could also undermine the effectiveness of so-called resistance-management refuges. Refuges are non-engineered crops planted in the proximity of engineered Bt crops to slow the evolution of insect pests resistant to the Bt toxin.

In theory, refuges work by allowing populations of susceptible pests to survive and mate with the relatively rare resistant pests. If the refuges are contaminated with Bt-producing plants, they would likely kill some susceptible pests, thereby aiding the emergence of Bt-resistant insects. Whether the presence of Bt transgenes in traditional varieties of crops would affect the efficacy of refuges would depend on the levels of contamination. Low levels of contamination would be unlikely to have much effect.

In general, seed contamination provides an avenue for release into the environment of genes and gene products that have not been evaluated or allowed in commerce and whose presence is unknown to farmers, regulators, or scientists. These may cause problems that are difficult to identify and remedy. Corn and soybeans are major export crops. The United States produces far more of these crops than its own economy can absorb, so it sells aggressively to the rest of the world.

The United States grows substantial quantities of non-engineered and organic products that face no customer resistance anywhere in the world and which, in many cases, even command premium prices. But much of the non-engineered grain and oilseed is contaminated with varying levels of genetic sequences derived from engineered varieties. This would not matter if U.S. export customers tolerated contamination with engineered sequences to the same degree they tolerate contamination with other varieties or even other crops, but that is not the case. Many customers want grain or oilseed free of transgenic sequences, especially genes or gene products that have not been approved in their countries.

Meeting this demand has proved a difficult challenge. Most of the contamination of bulk grain and oilseed products is the result of physical mixing that occurs routinely within the infrastructure of trucks, ships, and grain elevators that moves commodity crops to market. In addition, the outcrossing of pollen from engineered plants into neighbouring fields is unavoidable. The existing commodity infrastructure was never intended to transport different segregated streams of grain and oilseed from farms to food and feed processors. As long as the United States grows substantial acreages of engineered crops and does not alter its commodity infrastructure, it will not be able to readily provide uncontaminated commodity grain or oilseed product.

Seed contamination exacerbates the difficulty of keeping engineered genetic sequences out of non-engineered grain and oilseed. Even if growers seeking to export highly pure non-engineered commodity crops could start with pure seed, unreviewed or unwanted transgenic sequences could move into their products via mixing or outcrossing. But, when farmers start with contaminated seed, even the most innovative systems for moving segregated products to market are doomed. Such systems represent new market opportunities and are currently the focus of substantial investments. It should also be noted that customer preferences are moving targets.

If international customers grow more accepting of engineered grain and oilseed, the intermixing inevitable within the current commodity system would cause fewer problems for U.S. exports, and the importance of seed contamination as a contributor to trade problems would be diminished. Global resistance to genetic engineering, on the other hand, could continue to stiffen and perhaps reach a point where the United States would have to retool parts of its commodity grain and oilseed infrastructure to enable the segregation of uncontaminated non-engineered products.

Organic Food Production

Food products that bear the federal organic seal and label have met the U.S. government's standards for the growing and handling of organic food. The core organic standards restrict the use of synthetic pesticides and prohibit the use of irradiation, municipal sludge, and engineered seeds and other engineered inputs in food production. The combination of comprehensive and stringent standards and management systems that enable farmers to meet these standards comprises a holistic approach to food production that works in concert with the environment.

Food that meets organic standards generally commands a premium price in the marketplace. In fact, organic food has sufficient appeal that it is one of the few sectors of U.S. agriculture that is maintaining long-term, double-digit annual growth rates. U.S. certified organic cropland and pasture more than doubled between 1992 and 2001, from fewer than one million acres in 1992 to 2.3 million acres in 2001. Because the organic market offers a value-added product especially important for small and medium-sized farms, the potential loss of this market is of growing importance to U.S. agriculture. Many organic buyers, processors, and consumers, like many U.S. export customers, are demanding a product free of transgenically derived sequences.

To the extent that U.S. organic farmers cannot meet that demand, consumers will go elsewhere or perhaps refuse to pay premium prices. The U.S. government, which touts its organic label as the equivalent of a label indicating the absence of genetically engineered sequences, also has an interest in helping organic growers meet the demand. Organic farmers are struggling to find uncontaminated seed. If they cannot purchase seed free of transgenically derived sequences or control post-planting outcrossing-neither of which is completely within their control — they will be unable to meet their own or larger societal demands for non-engineered food. Although it is only one part of the solution, the availability of seed free of engineered substances is essential to meeting consumer demand and preserving an increasingly important sector of U.S. agriculture.

The pervasive contamination of seeds may also have patent implications for farmers who inadvertently plant and harvest seed containing transgenically derived sequences. Under U.S. intellectual property laws, genes, gene products, and engineered crops are now considered patentable subject matter-just like windshield wipers or clocks. Where patents apply, it is illegal for others to make, use, or sell the invention during the term of the patent without permission from the patent holder. To do so could subject the infringer to lawsuits and stiff penalties. An important feature of the patent law is that infringement does not require intent.

Farmers who use genes or seeds patented by others can be sued even if they did not know they were using the invention. While the law is murky, pervasive seed contamination would appear to put farmers at risk of unknowingly infringing the patents held by biotechnology companies. The threat of patent holders pursuing infringement claims against farmers who inadvertently purchased contaminated seed seems counterintuitive, but it is not impossible. Monsanto, for example, has not been shy about bringing suits against farmers for patent infringement, despite having provoked widespread anger and resentment in rural America.

The presence of unapproved genes and gene products would of course play havoc with the food system if the traits they confer proved to be harmful, but this could be the case even if they were not harmful. In general, food handlers and processors are not allowed to sell food considered to be adulterated under the provisions of the Federal Food, Drug and Cosmetic Act (FFDCA). Food may be considered adulterated for many reasons, including the presence of either pesticidal substances for which the government has not set a tolerance or unapproved food additives. Products of genetic engineering could fall in either category.

The Environmental Protection Agency (EPA) had approved StarLink (a variety of corn engineered to contain a pesticidal Bt toxin) for animal feed but not human food in 1997. The announcement that StarLink corn had been found in taco shells in 2000 set into motion widespread product recalls. Without a tolerance set by the EPA, the presence of the Bt pesticide rendered food adulterated under the FFDCA and therefore illegal. The recalls set off an expensive chain of events as grain sellers and food handlers had to test and divert contaminated lots of grain.

The sources of transgenes that may move into the food supply and trigger similar disruptive events include food crops grown for non-food purposes and engineered crop varieties in the early stages of development prior to commercialisation. Transgenes from these sources might be physically mixed with or outcross into food crops destined for the food system, where they could cause widespread disruption including recalls and lawsuits if discovered. Seed contamination would exacerbate this problem by making it even more difficult for growers and food companies to know the exact composition of the products they buy and sell.

Agriculture in Developing Countries

There are two ways that seeds contaminated with engineered sequences could make their way to developing countries: as seeds for planting or as bulk products, which are made up of viable seeds. In developing countries, it is highly likely that seeds purchased as commodity products will be planted by farmers as seeds. Unsuspecting purchasers of potentially contaminated traditional seed in developing countries will take no precautions to prevent the flow of transgenes into nearby crops and wild and weedy relatives via outcrossing. Since U.S. seeds could be contaminated with many kinds of genes, the consequences of gene movement are difficult to predict.

Transgenes that confer fitness benefits on plants can become fixed in plant populations and increase in frequency in successive generations. Thus, seed contamination could become a conduit for new genes-some of which may be harmful to human health or the environment-into wild and weedy plants. In general, the most unsettling aspect of seed contamination for producers in the developing world is that there is no way to evaluate, monitor, or avoid such movements because they would occur surreptitiously. Where transgenes move into other varieties or landraces, they could lead to unpleasant-and expensive-surprises. For example, if herbicide-resistance genes move into crop varieties, farmers may find that costly herbicides do not work.

Or, if seeds of traditional corn varieties are contaminated with the Bt toxin gene, farmers may find that the crop unexpectedly kills beneficial insects. The impact of contaminated seed must be considered against the backdrop of genetically engineered varieties that enter a country legally and fully disclosed as a bulk commodity product. It is likely, for example, that Bt transgenes in Bt crop varieties diverted for use as seed would flow through pollen into neighbouring crop varieties, landraces, and wild relatives. For popular transgenes such as Bt, the seed diverted from transgenic varieties of commodity crops is likely to be a greater source of novel genes in developing countries than those same transgenes occasionally contaminating seeds of traditional varieties.

The 2001 discovery that landraces of corn in Mexico are contaminated with genetic sequences that originated in engineered corn varieties from the United States underscores the difficulty of confining transgenes used in agriculture. It is not clear how the genes traveled to Mexico-whether seeds unapproved in Mexico were sold on the black market or bulk products imported from the United States were diverted and used as seed. The Mexican government is attempting to assess the causes and consequences of this finding. The unexpectedly rapid dispersal of transgenes to Mexico only a few years after their first commercial use in the United States deserves immediate attention from the scientific community because Mexico is the centre of diversity for corn, one of the world's most important food crops.

Teosinte, the crop's wild progenitor, can be found growing in Mexican cornfields, and whatever novel genes are found in Mexican landraces are also likely to be transferred into teosinte plants via pollen. While it is impossible with our current level of knowledge to assess the impact of novel genes on teosinte populations, the potential contamination of such important populations of wild plants points to the need for additional research. The ongoing situation in Mexico highlights the ease with which novel genes and traits can move through agricultural varieties into wild plant populations, including the vital populations that are the centres of diversity for important crops.

Seed Repositories

Ongoing contamination of the commercial seed supply could gradually undermine the quality of the communal genetic storehouse for agricultural crops. Nothing is more fundamental to the future of our agriculture and food system than a continued supply of safe, highquality seed. The prowess of genetic engineers notwithstanding, seeds cannot be made from scratch. They must be produced generation after generation through highly complex, natural biological processes. The value to food and fiber production embodied in the seeds entrusted to the generation cannot be overstated. Plant genetic storehouses are maintained through dynamic processes that involve saving, selecting, and storing seeds.

A number of groups and institutions are involved in this process. First commercial seed companies develop and sell seeds for crop varieties destined for fields or home gardens, in some cases in cooperation with farmers, gardeners, and scientists. Public-sector plant breeders also develop new varieties, although their role has diminished over the last several decades. Farmers also continue to be active in seed selection and preservation. In fact, most of the world's farmers do not have access to commercial seed products and save seeds every season for planting the next season. In addition to seed stores that are actively managed by companies, scientists, and farmers, some seeds are gathered and kept in repositories called seed banks.

Some of the most important seed banks house collections managed by international organisations such as the Consultative Group on International Agricultural Research. These seeds contain genes for valuable traits and combinations of traits that have been selected in a process spanning countless generations. Seed banks are not static collections; the seeds are often removed and planted, and their progeny returned to the seed bank. Although motley and uneven in its importance to different farmers and gardeners, the sprawling network of seed repositories is vital to the quality and resilience of the food supply.

Seed bank officials moved immediately to recall contaminated seed samples that had been sent to researchers in the United States and 14 other countries since 1996. The magnitude of the threat posed by transgenically derived sequences is not known at this point. Engineered traits per se are not necessarily a problem from either are not necessarily a problem from either an environmental or human health standpoint. Agricultural scientists do not knowingly create harmful varieties (other than, perhaps, for pharm and industrial crops.) Eventually, if engineered varieties are used for a long period without ill effect, seeds from certain engineered varieties will likely be added to seed collections intentionally.

10

Protection of Food Supply

Pharmaceutical uses encompass the production of therapeutic agents for humans and animals, including drugs and vaccines. Industrial uses range widely from enzymes to plastics to lubricants. The benefits of these products, especially new therapies, could be considerable. On the other hand, the products may pose serious health, environmental, and commercial risks. None of these substances is intended for food use and many of them, especially pharmaceuticals, may be bioactive at very low concentrations.

Many pharma and industrial products will be potentially toxic substances that could harm humans, food animals, or wildlife if ingested in active forms. While there are differences between crops genetically engineered for pharmaceutical and industrial uses, they pose similar risks to consumers and the food industry because they all produce substances not intended for food use that may inadvertently end up in food.

There are two major routes by which pharmaceutical and industrial transgenes can inadvertently contaminate commodity crops, and therefore the food and feed supply. One is the physical mixing of seed-pharma seed can be inadvertently spilled or mixed during seed production, harvest, storage, transport, and handling. The other is pollen, which contains the male reproductive cells necessary for the fertilisation of plants and the production of seed.

Corn and Soybean

Corn and soybean-both major food crops grown on about 27 to 28 million hectares (about 70 million acres) each year-are currently being engineered to produce pharma and industrial compounds. The characteristics that make corn and soybean dominant grain commodities also make them ideal for use as pharma crops. Both are inexpensively and reliably produced, reducing pharma production risks. Seeds of both crops can be stored inexpensively for relatively long periods of time (corn more so than soybean), so the pharma product can be quickly retrieved from storage and purified on demand.

Both seeds are relatively simple chemically, making the isolation of pharma products from grain economically and technically feasible. In addition, both crops are technically amenable to genetic engineering, so in principle nearly any pharma product can be produced in either of these crops. Neither has a close, sexually compatible wild or weedy relative in the United States to which pharma genes could escape by gene flow. From an exclusive production perspective, therefore, corn and soybean are ideal for pharma production.

However, the U.S. commodity corn and soybean production systems are structured to mix grain from many sources before it is ultimately used. Because of this, without substantial modification, the present commodity system will not be able to keep the human food and animal feed supply systems distinct from each other. Moreover, once a transgene enters the commodity system it may be difficult to remove it, as illustrated by the continuing occurrence of StarLink corn in the human food supply. This problem raises the fundamental dilemma associated with pharma crops. The compounds produced by genetically engineered pharma plants are expected to lead to useful products that would have beneficial effects on human or animal health.

At the same time, these compounds can contaminate the food supply and the environment and may have detrimental health effects on humans or animals, putting food companies at risk for market losses, legal liability, and brand damage. Seed is used for the production of commodity or IP grains and is grown and handled where traceability and quality standards are part of the system. However, these quality standards are insufficient to ensure protection of the food and feed supply from pharma product contamination. Complicating matters further, there are ranges of separation and segregation within each of these systems.

To address the problem, it is essential to understand the corn and soybean production systems as value chains-differentiated supply chains formed by a string of companies or collaborating operations that work together to produce specific products or services. These value chains would likely include transgene suppliers, seed companies, farmers, buyers and handlers of grain, and ultimately end users. A value chain of collaborating business entities requires cooperation, usually governed by contractual arrangements between the various parties in the chain. Without the full cooperation of all participants, it will be difficult to find a solution to the problem.

To evaluate the prospects for preventing contamination of the food supply by pharmaceutical transgenes, we must first identify the routes by which such contamination might occur. The two major routes of contamination (i.e., the presence of pharma transgenes where they are not wanted) are through pollen and seed. Pollen is essential for sexual reproduction and seed formation in corn and soybean. Eliminating the dispersal of pollen is one of the most straightforward ways to prevent biological contamination at a very early stage.

Seeds can be dispersed either naturally, by wind or animals, or by human handling through spillage during seed production, harvest, transportation, and storage. In addition, seeds can be mixed with other seed stocks during these same operations. Seeds spilled during harvesting, for example, can give rise to plants called volunteers in the next growing season. Grain residues left in a harvester can lead to the mixing of grain from different sources of the same crop.

Dispersal of Pollen

Pollen is dispersed mainly by wind and animals. For example, after an insect visits a flower in one population, pollen grains from that flower can stick to its body and be deposited in a flower of a different population. Pollen dispersal is a natural and widespread phenomenon; at least 12 of the 13 most important food crops in the world cross with wild relatives. Pollination involving transgenes may move genes for new traits, including pharma traits, from one plant to another. Unlike synthetic pesticides such as DDT or parathion, which are ultimately degraded in the environment, genes (including transgenes) are replicating molecules. They are able to maintain themselves in living organisms, where they become subject to evolutionary forces such as migration and selection.

In some cases, they become extinct in the environment, but in others where, for example, they provide a fitness benefit, their frequency may increase and they may spread in wild plant populations. Pollen can carry transgenes between crop varieties or between crops and wild or domesticated relatives. This kind of crossing usually leads to viable and fertile offspring that often display more vigorous growth than either parent. If pharma transgenes are transferred to and maintain themselves in populations of wild relatives, they could ultimately be reintroduced into non-pharma crops. Various factors influence the magnitude and geographic extent of pollen movement, including the reproductive mode of the plant, the topography of the surroundings (including the distances among populations), biological and physical characteristics of pollen, environmental factors such as climate and weather, and factors affecting the success of hybridisation.

Among the latter are the sexual compatibility between parents, presence cf compatible plants within pollination distance, sufficient overlap in flowering time, and the actual occurrence of fertilisation. The biological features of corn (Zea mays) and and mays soybean (Glycine max) differ from one another in differ from one another in max ways that affect the likelihood of genes escaping from the crops. The major difference between the two is that corn is a wind-pollinated plant whose offspring result from unions of pollen and eggs that come from different corn plants (outcrossing), while soybean tends to produce offspring that are the result of unions between pollen and eggs from the same plant (selfing), with only occasional outcrossing involving insect pollinators.

Corn Reproduction

Corn is a species with separate female and male flowers on the same individual plant. The female flowers are found in the ear on the side of the main stem, whereas the male flowers, called anthers, are located in the tassel at the top of the plant. Each flower in the ear consists of an ovule, style, and stigma. The ovules contain the eggs, which are fertilised by pollen grains. The styles, also called silks, are pollinated when they emerge from the ear's husks. The stigma is the receptive part of the style to which pollen adheres. An ear has approximately 1,000 ovules. By contrast, anthers in an average tassel will produce approximately 4.3 million to 5.2 million pollen grains. Thus, for each ovule, corn plants produce some 5,000 pollen grains.

The large number of pollen grains is characteristic of windpollinated plants. Two features prevent the release of seeds from a corn ear at maturity as happens in wild corn in Mexico. First, the ear is surrounded by husks, and, second, the ear does not break up to free the seeds. In corn, male and female flower parts mature at different times to avoid self-pollination. Silks appear one to three days after the anthers on the same plant have started shedding their pollen, and become receptive to pollen when they emerge from the husks. It may take up to five or six days for all silks to emerge from the husks. Each silk may remain receptive for up to 10 days, but high temperature and low humidity decrease this period.

Pollen shedding usually starts three hours after daybreak and continues for one to three hours. Under cooler temperatures and higher humidity, pollen dispersal may be delayed until noon and continue well into the afternoon. Individual pollen grains remain viable for two hours at the most. The total period of pollen shedding from a tassel can vary from one to two days to more than a week. So, the actual pollen dispersal time and duration of a given variety are variable and unpredictable because they depend on weather conditions. Whether dispersed pollen will be able to effect pollination depends on the presence of receptive silks, which, as noted above, is dependent on weather conditions.

Corn pollen is dispersed by wind over rather short distances because individual grains are heavy and large compared with pollen grains of other species. The dispersal distance is positively correlated with the size of the source field and wind speed. Other factors that influence pollen movement are gravity, wind direction, turbulence, air density and viscosity, pollen radius, and sedimentation velocity.

Soybean Reproduction

Soybean is a predominantly self-pollinating species with very low outcrossing rates. In domesticated soybean, outcrossing rates of only 1 and 0.5 percent have been reported for adjacent plants within rows and between plants of adjacent rows, respectively. Ahrent and Caviness and Nakayama and Yamaguchi reported a mean of 0.7 percent outcrossing in domesticated soybean, with a range of 0 to 5.89 percent. Soybean is a highly self-pollinated plant in large part because anthers and stigma are close to each other in the flower. (Unlike corn, single flowers of soybean contain both female and male organs.) The stigma is receptive to pollen one day before flower opening and remains receptive for two days after flower opening.

In most flowers, the anthers open and shed their pollen directly on the stigma on the day the flower opens. Under certain circumstances, such as cooler weather, pollen shedding and fertilisation take place within closed flowers. On average, each of the 10 anthers produces 300 to 800 pollen grains. Within 10 hours of flower opening, pollen germinates and fertilises the egg in the ovule. Soybeans may also be cross-pollinated by insects. With nectar consisting of 30 to 50 percent dissolved solids, soybean flowers attract insects such as honeybees, bumblebees, leafcutter bees, Halictid bees, and thrips.

Genotypic differences in outcrossing among varieties and higher levels of outcrossing in wild soybean suggest a partly genetic control in addition to environmental effects, principally the level of pollinating insects. Thus, although soybean is predominantly selfing, insect pollination is a mode by which pharma transgenes will escape, if pollintors are present. In summary, pharma transgenes may escape via pollen in both corn and soybean, despite their different flower biologies.

Seed Dispersal and Mixing

Seeds can also act as gene dispersal agents in two ways: they can be dispersed naturally-by wind, animals, or humans-or they can be spilled during seed production, harvest, storage, and transportation, which may cause future contamination of crops. These seeds may then grow into plants (volunteers) in future crops or may grow in noncrop areas, which could lead to pollen flow. Other plant parts could be harvested with future crops if not completely covered in tillage operations. Seed mixing is the inadvertent introduction of unwanted seeds or grains into a seed or grain product during the growing, harvesting, handling, storage, delivery, and processing of that product.

In the broadest terms, seed mixing, as applied to grain, is the mixing of two different types (species) of grains, which, in grain industry terms, affects "other crop" purity. An example would be corn grains in a soybean crop sample. A more specific type would be mixing seeds of two different varieties within the same species, which, in seed industry terms, affects "genetic" purity. An example would be the presence of a small amount of variety B soybeans in a sample of variety A soybeans. In commercial seed, the offending contaminant might not be identified, other than as "off-type."

Seed mixing occurs in both corn and soybean systems and for the same reason: the cropping methods and the equipment used for growing, harvesting, handling, and storage are very similar. The major difference between the two crops is the process of drying the corn grain after harvest. Soybean seeds usually reach storable moisture levels in the field, so there is no need for artificial drying. Corn, however, is harvested in a large part of its production area at moisture levels above those needed for proper storage, and requires drying with air in a grain dryer or in aerated storage bins. Seeds of corn and soybean, stored under dry (less than 10 percent relative humidity) and cool (less than 10°C) conditions, remain viable for several years.

Currently, most of the sampling and testing of the seed mixtures described above involves obtaining a representative sample, making a visual inspection, or conducting more elaborate testing procedures involving enzyme or DNA analyses. Both the grain and seed industries provide tolerances for these types of mixtures. However, these were established to maintain seed purity and are less stringent than a virtually zero contamination standard. Mixing of pharma crop and food crop seeds may occur at a number of points in seed production and in the grain production/handling/storage system-during the operation of equipment or the transfer of seed or crops between steps in the system.

For example, pharma seed may lodge in equipment or facilities and not be found during the cleaning and inspection between pharma crop and food crop operations. Pharma seed remaining in a combine after harvesting could contaminate the next product harvested with this equipment. At a storage facility, grain spilled during dumping from transportation equipment into storage or processing facilities could become mixed with grain handled later in the same facility. A number of confinement options are available to block the routes of contamination in corn and soybean systems; they can broadly be classified as follows:

— *Zoning*: Restricting the growth of pharma crops to areas of the country where corn and soybean are not usually produced

— *Spatial separation*: Growing pharma crops in fields separated from conventional crops by distances far enough that cross-pollination is unlikely

— *Temporal separation*: Planting pharma and conventional crops at different times to prevent overlapping flowering periods

— *Dedicated machinery/equipment/ infrastructure*: Reserving farm machines and other infrastructure for use with pharma crops exclusively

— *Physical confinement*: indoor production Growing pharma crops within physical structures

— *Biological confinement*: Employing various biological methods of reducing pollen or seed dispersal such as chloroplast engineering, male sterility, emasculation, seed sterility, cleistogamy, and apomixis

— *Disallowing food/feed crops*: Prohibiting the use of crops used for food or feed as pharma crops

Below we describe each of these confinement options and note their advantages and disadvantages when used as a single confinement measure within corn and soybean systems.

Zoning

Zoning is a confinement option that would restrict the growth of a transgenic pharma crop to areas outside the main production areas of the non-pharma food or feed crop. This option would mainly block contamination due to pollen dispersal but could also be implemented to reduce physical mixing. Corn is grown principally in the eastern half of the United States (east of 104° west longitude) with two main exceptions: the central valley of California and a region where the borders of Idaho, Oregon, and Washington meet.

In addition, there are areas in southern Arizona and New Mexico where white corn is grown for export to Mexico and other corn is grown for silage. American Indians throughout this western area may also grow corn on a small scale. If a zoning option were in place, producers in areas outside these production regions would be allowed to grow pharma corn. Some companies have already used this approach by locating pharma corn trials in western states. The main growing region for soybean is similar to that for corn, except that soybean is not grown on a significant scale in the western states.

This approach virtually eliminates the escape of pharma transgenes via dispersal of pharma crop pollen. It would also virtually eliminate physical mixing provided there were neither other fields of the same crop nor stands of wild relatives growing in the vicinity of the pharma crop. Wild populations of a pharma crop could theoretically act as a relay population between successive growing seasons of the pharma crop. Gene flow in both directions (pharma to wild and wild to pharma) could maintain the pharma gene in a certain area and could also lead to "stacking" of multiple pharma genes. This appears to be unlikely for corn, as the nearest wild-growing relatives are scattered populations in northern Mexico, and is irrelevant for soybean given the absence of wild relatives in North America.

Although not common in such areas, corn and soybean may also be grown outside of the major production areas. Pollen dispersal to those crops could be addressed if spatial separation complemented the zoning requirement. The zoning approach would allow the continued use of well-known crops and cropping systems, reducing or eliminating the need to conduct research on cropping methods for new pharma crops. Except for the challenge of identifying and establishing cultivation zones, it is readily applicable and entails little additional cost except for the higher shipping costs. If combined with other measures, such as dedicated machinery and infrastructure, zoning may also address contamination by seed mixing. It can be easily overlaid with existing management approaches, including identity preservation.

A major disadvantage of the zoning option is that the absence of other fields of the same crop within isolating distances would need to be verified before the planting of the pharma crop and would need to be monitored until the flowering period of the pharma crop ends. A permitting process tied to an Internet-based geographic information system might be needed to assist in siting pharma crop fields under a zoning option. Such a system would have to be designed to function as a central source of information enabling growers to identify plots located at a sufficient distance to ensure spatial separation complementing the zoning isolation.

In addition, implementing this option would require the enactment of legal restrictions excluding pharma crops from some regions and, perhaps, setting aside zones for the exclusive cultivation of pharma crops in other regions. A zoning system may also be difficult to set up in practice because of the non-commercial production of food or feed crops within the zone. Certain crops such as corn, for example, are grown by American Indian communities in the Southwest not to sell but as part of their heritage. Furthermore, corn is also grown in home gardens. These situations would require careful monitoring and perhaps agreements with neighbors. Another difficulty is the lack of optimal growing conditions outside the conventional growing regions.

Pharma crop growers would need to find areas within permitted zones that provide appropriate soil and climate conditions for the crop involved, including proper length and timing of the growing period, water availability, and temperature. In areas of the country where corn and soybean have not traditionally been grown, such as the desert west, other facilities including irrigation, adequate and dedicated transportation, and seed storage facilities would have to be developed. Thus, a zoning system may be difficult to set up in practice. Moreover, even if successful, implementation of this approach may soon exhaust the supply of land that satisfies the prerequisites for its application, depending on the scale at which the pharma crop is produced.

Spatial separation

Spatial separation is a form of confinement that allows pharma crops to be grown in the major crop-growing areas but locates them far enough from food and feed crops to significantly reduce pollen flow between crops. This option reduces contamination via pollen dispersal but has no impact on physical seed mixing.

In wind-pollinated plants such as corn, Tauber recognised three components in pollen dispersal, each associated with a different dispersal distance. The first is gravity, which acts over a very short distance. Basically, if gravity is the only process at work, pollen is deposited only on the parent plant or adjacent plants. The second, the local pollen component, depends on wind and shows a steep gradient in concentrations with distance downwind. This component is most likely to be affected by vegetation and other local physical barriers. The third, the regional component, consists of pollen grains caught by upward air movements and transported by airflows above the height of vegetation. and the local air current conditions created by surface features.

Regional airflows may carry pollen considerable distances downwind. The relative amounts of pollen dispersed by these three basic mechanisms will differ with factors such as the type of plant, the prevailing weather, and the time of day of pollen release. Unlike seeds, however, pollen grains are short-lived (for example, corn pollen typically survives for less than two hours). This places a limit on the regionalscale distances over which cross-pollination can occur. Factors that increase the dispersal of pollen include: a large plant population as the pollen source; the crop's reproductive system (a windpollinated plant such as corn often produces a high number of pollen grains, which are dry and airborne as opposed to predominantly selfing plants such as soybean, which have generally fewer, heavier pollen grains); the weather (sunny and dry days, strong winds); and the local environment (the absence of local vegetation and crop barriers, the presence of insect pollinators).

Corn pollen has been observed at 800 meters (0.5 mile) from the source as measured by an outcrossing frequency of 0.2 percent. The most efficient physical barriers are hedges and woodlands, which are penetrable to air flow, as these encourage the removal of pollen from the wind by filtration and minimise the formation of strong downdrafts in the lee of the barriers, which can concentrate the pollen at ground level. Similarly, border rows on both the source and the receptor crops can be efficient at minimising cross-pollination. It may be possible to reduce soybean cross-pollination with appropriate border plants. Although long-range pollen dispersal is a small fraction of total dispersal, it is a regular feature of plant reproduction and causes concern when pollen grains carry pharma genes. This is especially the case for corn, which is wind-pollinated and produces a very large amount of pollen per plant.

Setting isolation distances to eliminate longdistance travel of corn pollen will require much larger separation distances than those commonly used in certified seed production-on the order of several miles. Those standards were typically established to meet seed purity standards, most of which accept a relatively higher tolerance of contamination by other varieties of crops or even other crops. For example, certified soybean seed may contain up to 1.1 percent seed of other crops or other soybean varieties.

A standard of virtually zero contamination of the food system does not tolerate the levels of contamination set for certi-fied seed production and needs far stricter standards. Adopting the spatial separation option for pharma corn grown in corn-producing states could minimise, but cannot eliminate, crosspollination even when combined with physical barriers and border rows. In these states, many miles of isolation would be needed around pharmaceutical-producing corn, including isolation from home gardens. Therefore, virtually zero contamination of the food supply by pharma corn cannot be achieved only by spatially separating pharma corn from food and feed corn in the major corn-producing states.

For soybean, the spatial isolation distances recommended for commodity seed production are one to three meters (3 to 10 feet) and are set primarily to prevent mechanical mixtures of adjacent varieties or crops in the field. They are not intended to prevent occasional insect-vectored cross-pollination. Although this type of crosspollination is rare in soybean, it can be a source of escape for crop genes, including pharma genes. Isolation distances to prevent this from occurring depend foremost on the maximum flight distances of insect pollinators. The maximum measured flight distances of honeybees include 6.5 to 13.7 kilometers (4 to 8.5 miles), 4.3 to 6.2 kilometers (2.7 to 3.8 miles), and 10 kilometers (6.2 miles).

The largest distances are observed in settings without competing food sources. For leafcutter bees, a maximum distance of one kilometer (0.6 mile) was observed from alfalfa fields. Further experiments on individual plants led St. Amand, Skinner, and Peaden to recom-mend an isolation distance of some 1.6 kilometers (one mile). Further research is needed to characterise pollinators and their flight distances in soybean. In addition, actual gene flow distances-which are typically shorter than pollinator distances-need to be characterised.

The spatial separation confinement option has several advantages. This approach allows production of pharma corn and soybean in areas where corn and soybean grow the best and where farmers are expert in growing the crops. At the farmer level, the cost of adopting spatial separation may be minimal. This option can be readily instituted except for the problems of determining, implementing, and monitoring separation distances. It may be more effective for soybean than corn, given the shorter isolation distances needed. Used alone, spatial separation cannot achieve virtually zero contamination in either corn or soybean. It can be scaled up, but that will create additional problems of farmer adherence and cooperation, as well as enforcement.

Temporal separation

Temporal separation is meant to prevent an overlap in flowering times between pharma and non-pharma crops. It is an option that would affect pollen dispersal of pharma transgenes but would have no impact on physical mixing or seed dispersal of transgenes. Implementing this option for corn requires information about other fields within pollinating distances, their planting dates, the varieties planted, the timing of male flower opening and silking, and the duration of male and female flowering for corn. For soybean, implementation requires information on a variety's sensitivity to day length and growth habit, both of which are important factors influencing the flowering time.

Furthermore, in order to predict flowering time and determine when to plant other corn or soybean crops within isolation distance, intense monitoring of the weather (particularly in terms of temperature) is needed up until the end of the flowering time. This option is difficult to implement because there is no control over the weather, which is a major factor in the timing of flowering in plants. For example, under warm springs and cool summers, two corn crops planted one month apart will overlap minimally in flowering, but under cool springs and hot summers, they could overlap substantially, depending on the specific planting time.

The main, and perhaps sole, advantage of this option is that it can be implemented immediately. This option also has several disadvantages. Used alone, temporal separation is ineffective in controlling pollen dispersal because it depends on the vagaries of year-to-year weather variation and the diversity of crop varieties. Each variety has its own characteristic growth pattern, which is also influenced by the weather, especially temperature and water availability.

Thus, predicting the actual flowering time and duration for each variety is exceedingly difficult. Furthermore, this option requires intense monitoring of fields in the region and knowledge of varieties grown by neighbors of pharma crop producers. Planting at dates other than the normal planting dates is disruptive to normal production schedules and often leads to yield decreases. Enforcement would be difficult because inspectors would be unable to observe whether or not farmers had achieved temporal separation (except during flowering periods).

Dedicated machinery/equipment/ infrastructure

This option, requiring that certain equipment, machinery, and infrastructure be reserved for handling pharma crops exclusively, provides excellent protection against contamination by physical seed mixing, but does not affect pollen dispersal. The descriptions below indicate which types of equipment, machinery, and infrastructure should be dedicated solely to pharma crops and which could also be used for other crops. For the latter, cleaning and other methods mentioned below should be sufficient to protect against mixing.

Planting equipment-planters or drills used to place seeds in the ground: Planting equipment must be dedicated to avoid seed mixing because planting errors are common, and an unrecognised planting error with pharma seeds could have disastrous consequences. Even though this equipment can usually be cleaned easily, the possibility remains that seeds could be missed. Since planting equipment is expensive, older and/or reconditioned machinery might be dedicated to small acreages of pharma crops.

Harvesting equipment-combines and other harvesting equipment used to remove the grain from plants in the field: Harvesting equipment must be dedicated to avoid seed mixing. It is the most difficult equipment to clean with assurance that all seeds have been removed. Since this is the most expensive equipment in the production system, with new equipment costing approximately $200,000, the same approach suggested above for planting equipment can be applied (i.e., older, reconditioned harvesting machinery might be dedicated to small acreages of pharma crops).

Handling and transporting equipment- implements used to move grain from harvesting equipment to a storage facility or to deliver grain to an end user: This equipment must be dedicated. Commodity grain production involves many handling and transporting steps, from harvesting the crop through its end use. Elaborate systems could include wagons, unloading augers, dump pits and elevator legs, conveyor belts, bins, and trucks. In addition, corn production would require a grain dryer using heated air and the equipment needed to move the grain in and out of the dryer. Since the best approach to minimising mixing is to eliminate as many of the handling steps as possible, the most efficient dedicated system would dump crops from the harvesting equipment (combine) into a truck or container that would go directly to the end user.

Storage equipment (between harvest and end use)-bins or other containers used to store the crop and the machinery used to move the crop into and out of storagel: Unless the crop is delivered directly from harvesting equipment to the end user, it would be stored either on the farm or by a handler until the end user needs it. Storage equipment must be dedicated. Storage containers include round grain bins with perforated aeration floors, hopper-bottom bins, and cement or glasslined silos.

Delivery equipment (to the end user)- hopper-bottomed trailer trucks, railcars, or closed containers: This equipment must be dedicated because commodity grain production involves many handling and transporting steps in the delivery system. The most efficient dedicated system would rely on a dedicated container system.

The advantage of a dedicated machinery/ infrastructure system is that it nearly eliminates the potential contamination of food and feed supplies via the equipment and facilities used in pharma crop production. Physical seed mixing from machinery, handling equipment, and storage and transfer equipment is probably very common in commercial grain production. A second advantage is that it could be adapted immediately to pharma crop production. Currently, there is a surplus of seed-conditioning facilities in many parts of the United States that might be used as either dedicated pharma crop seed-conditioning facilities or dedicated commercial pharma crop-handling facilities.

For example, the Nebraska Foundation Seed Division (NFSD) has used a partially dedicated system for many years. Its growing, seed-conditioning, and storage facilities provide unusual opportunities in that they are located on a former World War II armament manufacturing and storage facility. NFSD has a lot of isolated land and many buildings that can be used for equipment and seed storage. It owns an older combine for each variety of soybean it will be producing, and that combine is dedicated to harvesting that one variety of soybean exclusively. NFSD may also use each combine for harvesting one variety of small grain that can be easily separated from soybean seeds. Other states may have similar, possibly underutilised handling systems that could be adapted to pharma crop production.

One disadvantage of this option is that it ties up capital investment in equipment and facilities. Some of this equipment is very expensive when purchased new and represents a major investment for the producer. Decisions on how to implement a dedicated-machinery option would depend on the size of the pharma crop operation relative to the total farming operation. For small acreages, older, smaller equipment could be dedicated to pharma crop production and newer, larger equipment to other production.

In larger-scale pharma crop production, the entire farming operation, including all the equipment, may be dedicated. Another disadvantage is the challenge of disposing of equipment and facilities dedicated to pharma crops. To allow equipment and facilities dedicated to pharma crops to be used for nonpharma crop production, it would be necessary to have certified equipment- and facility-cleaning processes.

Physical confinement

Physical confinement of pharma crops suffi- cient to block pollen dispersal can be accomplished using indoor production systems, including opaque structures that require artificial lighting and temperature maintenance, greenhouses with translucent or transparent ceiling and wall panels (with or without supplemental lighting), and intermediate structures using a mixture of artificial and natural light. Of these indoor systems, greenhouses are the best known-they are widely used around the world and the technology is well developed. They allow plants to be grown under close observation from seed production to planting to harvest, essentially completely isolated from other plants.

With temperature control, plants can complete their life cycle within a relatively well-defined period, and yields are usually more predictable than when crops are grown under rain-fed conditions out in the open. Water and nutrients are typically provided in indoor systems on an as-needed basis, and thus tend to be non-limiting. Adding carbon dioxide can increase seed yields of soybean but has less effect on corn. Corn, however, is a very lightresponsive crop, and maintaining light intensities equivalent to those in an outdoor field requires transparent panels and/or supplemental lighting.

Management of pests and diseases in enclosed systems is usually accomplished by chemical, cultural, and biological methods. Airlocks and decontamination procedures upon entry and exit can prevent pest or pollen movement into or out of contained facilities. Electronic alarm and control systems can be used to help maintain isolation. Indoor production facilities such as greenhouses are expensive, but the use of temporary, movable structures to isolate field-grown pharma corn during pollination would lower costs. Such structures are not in wide use now, but producing them using existing greenhouse technology is feasible.

Devising ways to shed heat loads from inside the structure will require additional development, but it should be possible to use such structures to seal corn plants from the outside for the two- to three-week period when there is a possibility of pollen release. Blocking pollen dispersal could be supplemented in corn by detasseling before the movable structures are lifted off the pharma crops. With plantings spread by one month or so, such structures could be used to isolate two crops in a season. Their use in soybean would also be possible, but because insect movement of pollen is the major dispersal mechanism in soybean, using mesh structures might be preferable to enclosures.

Physical confinement of crops within structures is a very effective means for preventing pollen movement. Yields of pharma crops grown in enclosed structures would be more predictable and less variable than in the outdoors, and it should be possible to produce two or three crops per year, even in temperate climates. This option could be implemented in the very short time needed to build and optimise the facilities.

Some disadvantages are;

— Enclosed structures are expensive to build and maintain.

— Indoor crop production systems require large amounts of energy to maintain temperatures within an acceptable range.

— Failure of electric power could mean rapid loss of a crop due to temperature spikes.

— Maintaining adequate light for maximum yields of corn is difficult and energy-expensive.

— Pest management is often difficult, especially when crops are grown to maturity.

— Windstorms and hail can destroy structures and compromise isolation.

— Though largely preventable through electronic alarm and control systems, human error can result in loss of isolation.

— Because they are highly visible, sabotage of structures, with loss of isolation, would be a constant threat.

— Personnel costs per unit of pharmaceutical production would in most cases be higher than if such crops were grown outdoors.

— The use of structures will necessarily limit the scale of production. cellular structures that are not carried by pollen.

Biological confinement

One approach to restricting transgene dispersal via pollen is to place the genes in the DNA of cellular structures that are not carried by pollen. One such structure is the chloroplast, which is responsible for turning light energy into chemical energy in plant cells. In most flowering plants, chloroplasts are inherited only through the female parent and are not carried in pollen grains. Transgenes inserted into the chloroplast DNA of most crop plants will not be carried by pollen to other plants.

If chloroplast genetic engineering were developed to the point that it could ensure transgenefree pollen, it would be an elegant way to avoid the dispersal of pharma transgenes to other crops or wild plants. So far, engineering chloroplast DNA to keep transgenes out of pollen has been successfully demonstrated in several crops including tobacco, tomato, cotton, and soybean. Although the chloroplast DNA of other crops, such as potato and carrot, has been engineered, the pollen of these plants has not been tested for the presence or absence of transgenes because of the difficulty in producing flowers under *in vitro* conditions.

More than 40 transgenes have been stably integrated into chloroplast DNA to confer desired plant traits or produce pharmaceuticals, edible vaccines, and industrial chemicals. In addition to its potential for confining pharma transgenes, chloroplast engineering offers other advantages over the more common nuclear engineering. For example, restricting transgenes to chloroplasts means that transgene products including drugs and edible vaccines will be concentrated in tissues such as leaves and fruits where chloroplasts predominate. It is also possible to engineer multiple genes into chloroplast DNA in a single transformation event and to use antibioticfree selectable markers or eliminate marker genes.

Chloroplast genetic engineering is also free of the position effects and gene silencing often observed in nuclear engineering, as well as harmful transgene products that cause pleiotropic effects. The NRC summarises the strengths of chloroplast engineering as follows: chloroplastspecific transgenes would not be spread by the pollen of most cultivated plants, and this approach could prevent transgene dispersal in pollen while preventing some of the disadvantages of male sterility (e.g., a loss of pollen for cross-pollination).

The disadvantages of chloroplast engineering are: the lack of success so far in moving transgenes into chloroplasts in some major crops, particularly corn and other cereal grains; the difficulty of establishing on a case-by-case basis and with certainty that pollen does not transmit engineered chloroplasts; and its inability to prevent transgene movement via seeds. While the efficiency of chloroplast engineering is high in some crops, such as tobacco, cotton, soybean, and carrot, studies are under way in a number of crops, including potato and tomato to improve efficiency and reliably exclude chloroplast DNA from pollen.

Male sterility

Male-sterile plants do not produce viable pollen capable of fertilising eggs from the same or another plant. Male sterility occurs naturally in some plants, including corn and soybean, but can also be engineered. In conventionally bred crops, cytoplasmic male sterility-sterility controlled by non-nuclear genes-is used extensively to facilitate hybrid seed production. Engineered nuclear male sterility is also used in plant breeding; one example is transgenic herbicide-tolerant oilseed rape, which comprises about 10 percent of the commercially cultivated crop in Canada. Although male sterility was at one time used in commercial corn production, it was abandoned because of susceptibility to southern corn leaf blight. Alternative sources of male sterility genes could be used that do not have negative traits associated with them.

Naturally occurring cytoplasmic male sterility already exists, is used in plant breeding, and could presumably be used to eliminate pollen dispersal of transgenes. Theoretically, the ability to engineer male sterility into crops, and therefore prevent ollen dispersal of transgenes, would allow a greater range of plants to be used for pharma production.

Confinement systems based on male sterility have several disadvantages. First, male-sterile systems may be leaky (i.e., they would not completely eliminate gene dispersal via pollen). For example, expression of naturally occurring male-sterile genes is known to be affected by weather conditions and the developmental stage of the plant. Leaky systems are sufficient for hybrid seed production but would not meet the virtually zero contamination standard for pharma crops in the food supply.

Second, neither naturally occurring cytoplasmic male sterility nor engineered nuclear male sterility has been developed for use in controlling transgene dispersal in major crops including corn and soybean. Third, while male sterility may be effective in preventing transgene dispersal in many cases, a crop engineered to be male-sterile could still be fertilised by pollen from wild relatives and serve as a female parent for hybrid seed. If such hybrid seed were to survive, germinate, grow, and reproduce, it would produce viable transgenic pollen that could cross-pollinate with weeds. Fourth, if seed is the desired end product of a transgenic male-sterile crop, cross-pollination with pollen from another source would be required. Finally, like chloroplast engineering, male sterility does not prevent seed dispersal of transgenes nor does it affect seed mixing.

Emasculation

Emasculation is the physical removal of male reproductive parts from a plant. As described earlier, corn and soybean have very different reproductive structures. Corn is easily emasculated to control pollen dispersal because male and female flowers are separated on the plant. Male reproductive structures are located in the tassel at the top of the plant, while the female reproductive structures are located in the ear on the side of the plant's main stem.

The emasculation, or detasseling, process is accomplished by pulling or cutting the tassel off prior to pollen shedding. This process has been used for decades in the corn seed industry to control pollination in the production of a hybrid from two different parental lines.

By contrast, soybean has totally enclosed flowers, containing both female and male parts. Emasculation in soybean is a time-consuming operation that is impractical on a commercial scale. Emasculation might be used in pharma corn production as follows: the pharma gene would be engineered in the female (ear) parent, pollination would occur by a non-pharma corn line, and the female parent would be detasseled (emasculated) to prevent dispersal of pharma pollen to nearby crops.

This method can eliminate pollen dispersal to surrounding corn crops if it is implemented effectively. Emasculation is not a practical method for soybean and most other crops because the male and female flower parts are too close together to allow easy physical removal of male parts. In corn, the detasseling process is laborintensive but used regularly in hybrid seed production. It must also be performed in a timely manner, regardless of weather conditions. Extreme weather, however, can compromise the process, allowing some pollen to escape (for example, if high winds blew through fields before detasseling was completed). Effective implementation would require a contingency plan and integration with other confinement measures such as spatial separation. Finally, emasculation, like chloroplast engineering and male sterility, does not prevent physical mixing or transgene dispersal via seeds.

Seed sterility

Seed sterility is a genetically engineered trait that leads to seeds that are formed normally on the plant but are incapable of germinating. Used as a confinement option for pharma crops, seed sterility would aim to reduce seed dispersal of pharma transgenes. Several seed sterility systems-variously called technology protection systems, gene use restriction technology, and "terminator" technology-have been proposed. So far, however, the approach has been successful only in tobacco. The lack of success reflects the fact that genetically engineered seed sterility systems are complex.

For example, the mechanism proposed originally by Russell, Hoopes, and Odell and Odell, Hoopes, and Vermerris consists of a three-transgene system. This system is triggered by a specific external stimulus such as the antibiotic tetracycline, which initiates interactions between two transgenes-interactions that ultimately unblock the third transgene. The third transgene codes for a toxin that kills seeds at a late stage in their development. As long as the third transgene is blocked, no toxin is produced. Once it is activated, however, it produces the toxin to prevent an otherwise fully developed seed from germinating.

A seed sterility mechanism successfully engineered into a pharma crop would, theoretically, substantially reduce dispersal of pharma transgenes via seeds. Seed sterility mechanisms have several disadvantages. They are complex and technically difficult to introduce into plants. In addition, the sterility trait may not function consistently. Two causes for failure might be the inactivation of one or more transgenes or the separation of one transgene from the other two.

Seed sterility systems are also not ready for use in pharma crop production-none of the proposed mechanisms have been successfully introduced into crops other than tobacco. Finally, this technology raises political, social, and ethical issues. For example, "terminator" technology is generally considered unacceptable in areas where farmers save seeds for planting in subsequent growing seasons, as is the case in many developing countries. Public access to data on the efficacy of transgenic reversible sterility will be essential. The technology should not be used in food crops for which growers need to save seeds for future planting or breeding. Possible environmental concerns should be evaluated on a case-by-case basis. This technology will not prevent clonal propagation of many plants, such as some species of grasses, shrubs, and trees.

Cleistogamy and apomixis

Cleistogamy is a process by which self-pollination and fertilisation occur within closed flowers. It has been suggested that crops engineered to exhibit cleistogamy could not disperse transgenes via pollen. Another option would be to engineer crops that produce seeds without fertilisation. This process, termed apomixis, occurs naturally in a small percentage of flowering plants. In apomictic plants, the embryo within the seed is not a product of sexual reproduction. Rather, it originates entirely from female flower parts. There are two types of apomictic plants. In one, the seed tissue feeding the embryo and new seedling results from fertilisation with pollen.

If genetically engineered apomictic crops resembled these plants, then apomixis would not alleviate the pollen dispersal concern as pollen is an integral part of the process. The second type of apomictic plants develops seed tissues without fertilisation. If crops were genetically engineered to mimic these plants, apomixis could theoretically block pollen dispersal. Although more than 400 apomictic plant species are known, apomixis is rare among crops and transfer of apomixis to crops by conventional breeding has been largely unsuccessful. Several methods for genetically engineering apomixis in crops have been proposed recently.

Cleistogamy, if successfully engineered into pharma crops, may be an effective barrier to pollen dispersal of transgenes. Engineering apomixis into crops would be advantageous in reducing pollen dispersal of transgenes if fertilisation were not required for seed production. Cleistogamy would appear to have limited efficacy, as it depends partially on uncontrollable and unpredictable environmental conditions. For example, in rice that exhibits cleistogamy, genes readily move between cultivated and wild forms of weedy rice, despite predominant self-pollination. Under cooler conditions (less than 20°C), soybean flowers remain closed and cleistogamy occurs.

When temperatures rise above this threshold, flowers on the same plant open and fertilisation occurs in an open flower. Thus, this option would not reduce pollen flow to virtually zero levels and does not reduce physical mixing. To date, very little work has focused on this approach and knowledge of genes responsible for flower development remains rudimentary. Any application is a long way off. In addition, transgene escape by seed or vegetative reproduction could still occur for plants with obligate cleistogamy. Engineering apomixis into crop plants may theoretically prevent pollen dispersal, but it would be effective only if pollen is not needed for seed development.

Apomixis is known to occur only in a few cultivated plants and, thus far, the trait has not been introduced into any transgenic crops, including corn and soybean. Studies of naturally occurring apomixis suggest this method could be leaky. Given that even obligate apomicts still produce seeds, the method cannot be used for confinement if the goal is to prevent dispersal by seed. Also, it will be important to confirm that apomictic transgenic crops cannot establish invasive populations.

Disallowing food/feed crops

Under this option, pharma genes could be engineered only

into crops that are not used for food or feed. This option precludes both corn and soybean. In the United States, corn and soybean are major feed crops and to a lesser extent food crops. (In other regions of the world, corn is principally a food crop.) The fact that corn and soybean are food and feed crops is the basis of concerns about pharma transgene dispersal by both pollen and seeds and physical mixing of pharma and food/ feed seeds.

A choice of non-food/feed crops would alleviate to a considerable extent those concerns and greatly facilitate the segregation of pharma and non-pharma crops. Other crops besides corn and soybean being tested as pharma crops include alfalfa, canola, potato, rice, safflower, and tobacco. All of these crops are ingested in some way, and with the exception of tobacco, all are either food or feed crops.

Use of non-food/feed pharma crops would completely block pollen dispersal and significantly reduce, but not eliminate, the risk of contamination by physical mixing into the food/feed system. For example, pharma seeds produced on stray (volunteer) plants that germinate and grow from pharma seeds left behind after the cultivation of the pharma crop may contaminate food or feed crops growing in that same field the next year. This option can be scaled up once the expression of pharma traits and agronomic procedures for non-food/feed crops have been worked out. This approach would be more easily enforced and monitored than those using a food/ feed plant to produce pharma products.

Except for tobacco, this option is not ready to implement. Restricting pharma crop production to non-food/feed crops would halt work on most, if not all, products currently under commercial development and would mean a loss to companies that have invested in food and feed crops. This approach requires identifying and developing suitable alternative plants, preferably crops, and implementing a breeding program to produce suitable varieties. Special field equipment to cultivate these crops may not be available or may need to be adapted from other crops.

Management of Pharma Corn and Soybean

The adoption of a comprehensive, structured management system focused on the potential pharma crop contamination of the food/feed system should be considered. Such management systems could either be adapted from identity preservation systems currently used in the seed and specialty crops industries or be designed from the ground up to prevent pharma crop contamination.

Identity preservation (IP) is a process or system of documenting the identity of a product and maintaining its segregation. An IP product has identifiable characteristics that have been maintained from the seed through all the steps of production and transportation to the end user. This strict production and delivery method uses the procedures of an effective internal segregation system, including observations, inspections, sampling, and testing, to ensure the presence (or absence) of certain traits. IP systems encompass activities and procedures as well as infrastructure, equipment, and technologies. Growers must follow strict growing and handling practices, including segregation, inspections, and cleaning of equipment to prevent other varieties from mixing with or contaminating the IP variety. Other parties that handle, transport, condition, or process the IP product must also maintain and document a similar segregation system.

The key to an IP system is traceability. Each production, processing, and delivery step is documented so products can be traced from the store shelf back to the farmers' fields and every stage in between. IP must include a system of verified steps following the crop through the entire production and delivery system. Testing crop samples as a stand-alone procedure does not qualify as an qualify as an IP system.

IP systems are established management systems used in the seed and specialty crop industries to maintain genetic purity of the product being produced. The companies and growers using these systems are well acquainted with the procedures and technologies required to maintain genetic purity within the crops being produced. IP systems have not yet been adapted to eliminate contamination of other crops and would need to be modified to be applied as a virtually zero contamination system.

However, IP systems could easily incorporate confinement management procedures to eliminate or reduce potential contamination of crops or related wild species by pollen flow or physical mixing during growing and handling. Even if modified to include effective confinement procedures, IP systems would still need to be used in conjunction with other confinement options to provide virtually zero contamination of food and feed crops. Most IP systems used in the seed and specialty crop industries tolerate some level of contamination depending on the requirements for a specific product.

Confinement Management Systems

These systems are not yet defined, but the following concept is in the early stages of development: forcing operators to document what processes are being used and how they are being conducted, then proving through records and audits that the processes, however described, are consistent. This system would not itself require specific or high-quality standards, but would incorporate several strategies for minimising any harmful effects on the environment or other plant materials caused by an organisation's activities.

Confinement management system standards would include requirements for a given organisation to follow in managing those processes that have an impact on the environment or other products. For example, contamination of the food/feed supply by pharma products from either pollen dispersal or physical mixing could potentially occur during pharma crop production because of human error or lack of planning. Hence, a formalised confinement management system would address this possibility for all parties involved, including plant breeders, seed multiplication entities, seed companies, and pharma crop producers, handlers, and users.

The confinement management system envisioned would be structured similar to an ISO 9001:2000 quality management system or an ISO 14001:1996 environmental management system. It is possible that a confinement management system could be developed from or included as part of either of these systems. Whether such a system is completely new or part of an established system, a third party would be involved in approving and overseeing the compliance of pharma crop business entities.

Requiring all parties to adhere to a formal confinement management system would tend to unify the management of pharma crops. The new system would include any government-mandated requirements in addition to its own structured confinement plans and would involve a third party outside the U.S. Department of Agriculture or the Food and Drug Administration. The uniformity provided by this system might be beneficial to those agencies in their assessments of pharma crop activities.

Formal management systems such as the ISO systems require a large commitment of time and dedication by management and employees to develop and implement. They are not "off-theshelf " systems; rather, they must be developed internally by the business entity, specific to its production processes and requirements. Developing a confinement management system as part of an already implemented quality or environmental management system would facilitate and shorten the process.

Industry-wide Marking Systems

Industry-wide adoption of a color scheme would facilitate the management of pharma crops. For example, bright pink might be designated as the marking color for pharma crop production systems exclusively, and would not be allowed in other agricultural uses. All planting seed could be delivered in bright pink bags or containers, and all dedicated equipment and facilities could be marked with bright pink paint at designated positions on the equipment. An equipment/facility inventory control system would track this equipment so it could not be used for other purposes without authorised "decontamination."

Only after decontamination could the color marking be removed from this equipment. This marking system could be extended to the pharma crops themselves. For example, a gene coding for a distinct color or morphological trait could be incorporated into the pharma varieties to clearly distinguish them from non-pharma varieties. Decisions of this type should be made by representatives of the seed and pharma crop industries, production agriculture, third-party consultants, and regulatory agencies, and should not be dictated only by government bodies.

11

Food Loss Prevention

Post-harvest losses of fruit and vegetables are more serious in devoloping countries then those in well developed countries. An additional constraint to improving this situation is that in moat developing countries the number of scientists concerned with post-harvest food losses is significantly lower than those involved in production research. In the early days of horticulture in wolf developed countries, heavy losses occurred in much the same manner as they do today in developing countries. Increasing industrialisation in technologically advanced nations gradually brought about improvements in crop handling.

Elaborate harvesting equipment replaced the crude harvesting tools. Collection centres were strategically established in major producing areas. Containers were remodelled to add more protection to the produce. Commercial storage plants were installed and grade standards adopted. Engineers and economists became more and more aware of raw material behaviour. Concomitant advances in Refrigeration Technology in the developed countries have made possible establishment of cold chains for the entire post-harvest and handling operations. At the institutional level post-harvest research was initiated.

Pilot packing houses were installed, coupled with the development of intensive training programmes. The improvement of product quality and reduction in post-harvest losses became the main concern of producers, middlemen, marketing specialists and consumers. Today, enormous volumes of quality horticultural crops produced in technologically advanced countries are made available to millions of people through improved post-harvest handling. Thus, historically and by necessity, post-harvest technology is part of the normal development processes in agriculture. These handling procedures are not fully recognised in less developed countries. Here agriculture may be characterised as disjointed.

Production is not linked with marketing. With perishable crops like fruits and vegetables, storage, packaging, transport and handling technologies are practically non-oxiatant, Hence, considerable amount a of, produce are loaf. Thus, as more fresh fruits and vegetables are needed to supply the growing population in less developed countries, as more produce is transported to non-producing areas, and as more commodities are stored longer to obtain a year-round supply, post-harvest lose prevention technology measures become paramount. It is distressing to note that so much time is being devoted to the culture of the plant, so much money spent on irrigation, fertilisation and crop protection measures only to be wasted about a week after harvest.

It is, therefore, important that post-harvest procedures be given as much attention as production practices the stages from planting until the product a roach the consuming public must be a mutual undertaking between the growers and those who will handle the products after harvest. Fruit a and vegetables, root a and tubers (horticultural crops) are quite different in nature from cereal grains and oilseeds the differences are summarised in Table 1. The causes of spoilage, the rate at which spoilage occurs, the degree of spoilage, and the actions needed to reduce spoilage are substantially different than for the cereals. Because of these differences it is necessary to design a different set of intervention programmes to reduce post-harvest losses in horticultural products.

Table 1. Comparison of Horticultural Crops

Cereals and Oilseeds	Horticultural Crops
Low moisture content, typically 10% to 20%	High moisture contort, typically 70% to 95%
Small unit size, typically less than 1 gram	Large unit size, typically 5 e to 5 kg
Very low respiration rate with very small generation of heat. Heat production is typically 0.05 megajoule/ton/day for dry grain	High to very high respiration rate. Heat production is typically 0.5 to 10 megajoule/ton/day at 0°C to 5 to 70 megajoule/ton/day at 20°C
Hard texture	Soft texture, easily bruised
Stable, natural shelf life is one to several years	Perishable, natural shelf life is a few days to several months
Losses usually caused by molds, insects and rodents sprouting, and bruising	Losses usually caused by rotting (bacteria, fungi), senescence
Losses in LDCs usually 10% to 20%	Losses in LDCs usually 15% to 50%

In the warm humid tropics, fresh fruit and vegetables have an extremely, low level of natural protection against the climate peat a and bio-chemical and physiological deterioration. It is also in the warm humid tropics that human diseases are moat numerous and widespread and consequently where the need for the nutritional value of fresh fruits and vegetables la moat essential. On the other hand, dried grains and legumes, and to a certain extent, root crops (yams), have a fair degree of natural protection against the various deteriorative elements, even though considerable post-harvest losses do occur.

Importance of Perishable Crops

In terms of importance as assessed by volume of production within the tropics these are cassava, potatoes, yams, sweet potatoes, taro and others. It is estimated that some 174 million tone of roots and tubers are produced annually of which cassava constitutes about 60%. Cassava is almost entirely produced and consumed in developing countries although in recent years there has developed a substantial trade in dehydrated cassava chips which are exported from developing countries to European countries as a low-goat animal feed ingredient.

The European Economic Community imported 2.3 million tons of dried cassava chips in 1975. Next in volume of production are potatoes, yams and sweet potatoes. These make contributions respectively about 14, 11 and 10%. At the end of the scale taro contributes a lowly 2.2% while miscellaneous tubers account for 2.5%.

The total annual production of horticultural crops in developing countries is 620 million tone which is 85% of the cereal production in these countries. This surpasses the relative world average whore horticultural crop production is only 70% of the cereal production. These high production figures indicate the important role that horticultural crops play in the economy of developing countries although it must be remembered that these crops are far higher in water content than are the cereals.

Horticultural crops are essential for a nutritionally balanced diet. Fruits and vegetables are the major source of vitamins A and C, a good source of calcium and iron, and they supply part of the requirements for a number of other minor nutrients. Roots, tubers, bananas and plantains are important sources of calories and also supply a number of minor nutrients, and some protein. In addition, horticultural crops add variety, enjoyment and a sense of satisfaction with the diet because of their appealing colours, flavours and textures.

For example, it has been said that although onions and garlic are not rich in nutrients, they make a vegetarian diet acceptable because of the savoury flavour they impart to the monotonous starchy diet in a developing country. It is no accident that onions rank third, and garlic ranks tenth in volume of vegetable production in developing countries. On all of these counts- economics, nutrition, acceptability- horticultural crops play a major role in developing countries, amply justifying the contortion that something should be done to reduce the high losses that presently occur in these commodities.

Post-harvest Food Loss

The post-harvest period ends when the food comes into the possession of the final consumer. Food moans weight of whole some edible material that would normally be consumed by humans, measured on a moisture-free best. Inedible portions such as skins, stalks, leaves, and seeds are not food. Potential foods (e.g., leaf protein) are not foods; they do not become food until they are accepted and consumed by large population. Feed (intended for consumption by animals) is not food.

The method of measuring the quantity of food in the post-harvest chain should be on the basis of weight expressed on a moisture-free basis. There will be times when information on losses in nutritional unite and economic losses will also be needed but these should not be the prime means of measuring post-harvest food losses. Loss means any change in the availability, edibility, wholesomeness or quality of the food that prevents it from being consumed by people.

Causes of losses

There are so many causes for .losses in the post-harvest food chain that it helps to classify them into 2 groups and a numb or of subgroups. Primary causes are those causes that directly affect the food. They may be classified into the following groups

a. *Biological*: Consumption of food by rodents, birds, monkeys and other large animals causes direct disappearance of food. Sometimes the level of contamination of food by the excreta, hair and feathers of animals and birds is so high that the food is condemned for human consumption. Insects cause both weight losses through consumption of the food and quality losses because of their frass, webbing, excreta, heating, and unpleasant odours that they can impart to food.

b. *Microbiological*: Damage to stored foods by fungi and bacteria. Microorganisms usually directly consume small amount e of the food but they damage the food to the point that it becomes unacceptable because of rotting or other defects. Toxic substances elaborated by molds (known as mycotoxins), cause some food to be condemned and hence lost. The beat known of the mycotoxins is aflatoxin (a liver carcinogen), which is produced by the mold Aspergillus flavus. Another mycotoxin which is found in some processed apple and pear products is patulin, which is formed in the apple by rotting organisms such as Penicillium expansum which infect fresh apples before they are processed.

c. *Chemical*: Many of the chemical constituents naturally present in stored foods spontaneously react causing lose of colour, flavour, texture and nutritional value. An example is the Maillard relation' that causes browning and dicolouration in dried fruits and other product a There can also be accidental or deliberate contamination of food with harmful chemicals such as pesticides or obnoxious chemicals ouch as lubricating oil

d. *Biochemical reactions*: A number of enzyme-activated reactions can occur in foods in storage giving rise to oft-flavours, discolouration and softening. One example of this problem is the unpleasant flavours that develop in frozen vegetables that have not been blanched to inactivate these enzymes before freezing.

e. *Mechanical*: Bruising, cutting' excessive pooling or trimming of horticultural products are causes of loss.

f. *Physical*: Excessive or insufficient heat or cold can spoil foods. Improper atmosphere in closely confined storage at times causes losses.

g. *Physiological*: Natural respiratory losses which occur in all living organisms account for a significant level of weight lose and moreover, the process generates heat. Changes which occur during ripening, senescence, including wilting, and termination of dormancy (e.g., sprouting) may increase the susceptibility of the commodity to mechanical damage or infection by pathogens. A reduction in nutritional level and consumer acceptance may also arise with these changes. Production of ethylene results in premature ripening of certain crops.

h. *Psychological*: Human aversion, such as "I don't fancy eating that today". In some cases food will not be eaten because of religious taboos.

Microbiological, mechanical and physiological factors cause moat of the losses in perishable crops.

Secondary causes are those that load to conditions that encourage a primary cause of loss. They are usually the result of inadequate or non-assistant capital expenditures, technology and quality control. Some examples are:

a. Inadequate harvesting, packaging and handling skills.

b. Lack of adequate containers for the transport and handling of perishables.

c. Storage facilities inadequate to protect the food.

d. Transportation inadequate to move the food to market before it spoils.

e. Inadequate refrigerated storage.

f. Inadequate drying equipment or poor drying season.

g. traditional processing and marketing systems can be responsible for high losses.

h. Legal standards can affect the retention or rejection of food for human use by being too lax or unduly strict.

i. Conscientious, knowledgeable management is essential for maintaining tool in good condition during marketing and storage.

j. Bumper crops can overload the post-harvest handling system or exceed the consumption need and cause excessive wastage.

Losses may occur anywhere from the point where the food has been harvested or gathered up to the point of consumption. For the sake of convenience the losses can be broken down into the following sub-headings:

a. Harvest. The separation of the commodity from the plant that produced it. In the case of roots, tubers and bulbs the commodity is lifted out of the soil.

b. Preparation. The preliminary separation or extraction of the edible from the non-edible portion, e.g., the peeling of fruits and vegetables.

c. Preservation is the prevention of lose and spoilage of foods. For example, the sun-drying of fruit, the use of refrigeration and the use of funficides to inhibit mold growth in fruits.

d. Processing is the conversion of edible food into another form more acceptable or more convenient to the consumer, for example, the manufacture of fruit juice and the canning of fruits and vegetables.

e. Storage is the holding of foods until consumption. Most storage is common storage (ambient temperature) but there are extensive storage capacities that can hold food under refrigerated or controlled atmosphere conditions.

f. Transportation. All forms of transportation are used to convey foods from the point of production to the ultimate point of consumption.

Post-harvest losses in fresh root/tuber crops have their origin in mechanical damage, physiological processes, infection by decay organisms and, occasionally, pest infestation. The losses caused by these processes may occur during all stages of the food supply system from crop maturity, through harvesting, transportation and storage. The degree of lose associated with these factors is determined by the plant material involved, the prevailing environmental conditions and management of the food supply system.

Magnitude of Losses

Reliable statistics on losses are few. It is possible to find individual cases with losses ranging from 0% to 100%. The extent of losses is highly variable depending on a number of conditions. Stable foods such as cereal grains can be stored in good condition for several years, whereas perishable foods such as fruit a and vegetables, spoil quickly unless given special treatment such as canning and freezing. The longer the time the food is stored the more opportunity there is for losses of all kinds to occur. Perishable crops generally suffer from higher losses than the cereals. A number of figures for the extent of loss is quoted in scientific literature and by the communications media, but much of this information is unreliable because the amount of loss has been estimated and has not been obtained by actual measurements.

There is often the temptation to cite "worst case" figures to dramatise the problem. Another problem is that even some of the figures that have been obtained by careful measurements are manipulated for various reasons. In some cases there is the temptation to exaggerate the figures of loss, particularly if there is a prospect that high figures of loss will prompt aid from donors. In other cases there is a temptation to minimise loss figures in order to prevent the embarrassment of acknowledging the magnitude of losses, or for political, financial or trading reasons.

Another precaution that needs to be taken in assessing overall losses is to ensure that the arithmetic of calculating loss figures is correct. The extent of losses can be unwittingly exaggerated unless the arithmetical calculations are correctly performed. Some examples of misleading arithmetical calculations are discussed by Bourne. The pattern of losses varies widely from country to country. There is a marked contrast between the site of major losses in developed countries and developing countries.

In a typical developed country losses may be fairly high during harvesting because the agricultural machinery that is used to harvest the crops leaves some of the commodity in the field and mechanically damages some of it. Considerable quantities of foods a may be discarded at the point of harvest because they are of the wrong size, shape or colour. These are planned losses. In developing countries harvest losses are usually lower because most of the crop is hand picked. The amount of material rejected in developing countries is less because the expectation of quality and uniformity is generally lower than for developed countries.

In developed countries losses are generally small during processing, storage and handling because of the efficiency of the equipment, good quality storage facilities, and close control of critical variables by a highly knowledgeable cadre of managers. In contrast, in developing countries losses in processing, storage and handling tend to be rather high because of poor facilities and frequently inadequate knowledge of methods to care for the food properly. The tragedy of these enormous losses in developing countries is not only that this is a severe economic loss in regions that are struggling to escape from poverty but also a major loss of important nutrients to population who are malnourished.

There appears to be no established generally accepted methodology for determining post-production lose in root/tuber crops. Assessment is a rough quantitative approximation of food loss or the characterisation of the relative points of lose in a particular food supply system. This approach implies a measure of subjectivity resulting from a lack of sufficient information. Measurement on the other hand is a more precise quantitative observation with less subjectivity. With measurements there is a high expectation of reproducibility without observer bias. Estimation is the interpretation of a number of scientific measurements. Here the process of interpretation depends on the experience and judgement of the observer.

At whatever level of precision post-harvest lose is determined the value will be specific in time and for location. This is due to the fact that lose is a function of the condition of the material, the prevailing environment, the nature and intensity of bio-degenerating organisms and the crop material management. None of these are constant. They are all dynamic factors liable to continuous change. As a consequence, crop loss, however determined will always be variable. This is illustrated by yam storage. Some cultivators of the rotundata/cayenensis/discorea species remain dormant for about 3 months, during which period storage loss is low.

When placed in the traditional yam barn the keeping characteristics of these yams are good. It is generally believed that little improvement can be achieved without radical change in storage technology. However where the tubers are infected by nematodes the storage potential is much reduced and high losses occur. It would be useful to have a standard method for assessing losses for each type of commodity but this is a difficult task due to crop diversity, inherent perishability and the complexity of the marketing and distribution channels.

Effects of the Environment

The environmental conditions under which foods are stored and processed can have a major effect on the keeping quality of the foods and the amount that is lost. The major environmental influences on the keeping quality of foods are the following:

Temperature: In general, the higher the temperature the shorter the storage life of horticultural products and the greater the amount of lose within a given time, as most factors that destroy the produce or lower its quality occur at a faster rate as the temperature increases.

Humidity: There is movement of water vapour between a food and its surrounding atmosphere in the direction towards equilibrium water activity in the food and the atmosphere, A moist food will give up moisture to the air while a dry food will Absorb moisture from the air. Fresh horticultural products have a high moisture content and need to be stored under conditions of high relative humidity in order to prevent moisture loss and wilting, exceptions to this being onions and garlic. Dried or dehydrated products need to be stored under conditions of low relative humidity in order to avoid absorbing moisture to the point where mold growth occurs.

Solar Radiation: The solar radiation that falls upon foods held in direct sunlight increases the temperature above the ambient temperature. The amount of increase in temperature depends on the intensity of the radiation, the size and shape of the food and the duration of exposure to the direct rays of the sun. The intensity of solar radiation depends upon latitude, altitude, season of the year, time of day, and degree of cloud cover. Under clear skies it is most intense when the sun is most directly overhead.

Hence the intensity of radiation is greater in tropical zones than in temperate zones. A high temperature is deleterious to food quality and increases wastage. It is ironic to note that, in the temperate climates where the intensity of solar radiation is moderate, almost all food is kept inside under cover whereas in tropical climates, where solar radiation is much higher, considerable quantities of food can often be found in the direct rays of the sun deteriorating in quality at a rapid rate.

Altitude: Within a given latitude the prevailing temperature is dependent upon the elevation when other factors are equal. There is on the average a drop in temperature of 6.5°C for each Km increase in elevation above sea level. Storing food at high altitudes will therefore tend to increase the storage life and decrease the losses in food provided it is kept out of the direct rays of the sun.

Atmosphere: The normal atmosphere contains by volume, approximately 78% nitrogen, 21% oxygen, 1% argon, 0.03% carbon dioxide various amounts of water vapour and traces of inert gases. Modifying the atmosphere can improve the shelf life and reduce wastage of certain foods. One type of controlled atmosphere storage (CA) is refrigerated storage in which the level of oxygen is reduced to about 3% with the carbon dioxide content being raised to 1 to 5%, depending on the commodity. This CA storage may double the storage life over that of regular cold storage for certain varieties of apples and pears by slowing down the natural rate of respiration. Many fruits, the "climacteric fruits", generate ethylene gas during ripening and the presence of this gas

If the ethylene is removed from the atmosphere surrounding these fruits as it is generated, their storage life may be extended. Experiments have shown that placing such fruits in a fairly gas-tight container with potassium permanganate, which absorbs ethylene gas, can substantially extend the storage life even at ambient temperature. "Modified atmosphere storage" is another type of controlled atmosphere storage. This term denotes storage of horticultural products in a beneficial atmosphere other than air that is not under as close regulation as in CA storage.

Modified atmosphere storage can be obtained in boxes of pears, apples, and cherries that are lined with polyethylene film which acts as a barrier to the escape of carbon dioxide and the ingress of oxygen. Another method of obtaining a modified atmosphere storage is by the addition of dry ice which increases the carbon dioxide in the atmosphere to some extent.

Time: The longer the time the food is stored the greater is the deterioration in quality and the greater is the chance of damage and loss. Hence storage time is a critical factor in lose of foods, especially those that have a short natural shelf life. The time involved from harvest to consumption of perishable commodities is much shorter than that from planting to harvest. While production time could take about several years for fruit trees, and generally about three to four months for vegetables, the duration of post-harvest handling could be as abort as one day to a few weeks. Any improvement in post-harvest handling treatments would therefore involve less risk and would be more economical than improvements in production.

Biological pressures: Bacteria and fungi are always present in the atmosphere to contaminate food and cause spoilage should conditions favour their growth. However, it should be emphasised that the contamination or inoculation process with bacteria and fungi occurs to an equal extent during the harvesting process. Soil organisms as well as foliage pathogens can be introduced. Bacteria that cause disease in plants are not usually introduced from the air par so except in aerosols. Micro-organisms can multiply very rapidly whenever conditions are favourable for growth. The only foods that are free from micro-organisms are those that have been thermally processed, such as canned goods.

A similar situation occurs with insects. Insects are in the field and can accompany foods as they are brought from the field into storage. Most of the stored products' insects can increase is number by a factor of 10 to 50 times par month under favourable conditions. Stored food insects are ubiquitous, hiding in storage facilities and moving with stored foods when they are moved. Consequently we can assume that foods can become infested with insect pests at any time unless special precautions are taken. Rodents and sometimes birds can exert biological pressures similar to those of insects and micro-organisms. The great capacity of living organisms for multiplication by geometric increase generates heavy biological pressures upon stored foods.

A number of people who have accepted the idea, of the need for better conservation of food wonder why the United Nations Environment Programme (UNEP) is interested in food losses and control of food losses. There are substantial reasons for UNEP's interest in this area and these will now be discussed. UNEP is interested in promoting the health and well-being of both people ea. the environment, as well as sustainable development. Reducing food losses should improve nutritional status and human health especially in those countries here a large proportion of the population is inadequately fed.

UNEP encourages the conservation of food because of its positive environmental effects. UNEP is interested in efficient and non-wasteful utilisation of resources. Although it is difficult to obtain firm figures, it is generally conceded that considerably less energy and other inputs are required to conserve food than to produce an equal quantity of food. For example, it has been estimated that the total energy cost of good grain storage practice is about one percent of the energy cost of producing that grain.

Hence UNEP encourages good conservation practice because of its more efficient use of energy and other inputs. UNEP's activities include reviewing proposed new initiatives to determine the environmental impact of these initiatives before they are put into effect and to help select from among competing initiatives those that are most desirable from the environment standpoint to encourage their use, and to do this in the early planning stages of the project. Hence, UNEP is interested in what methods are proposed to be used to reduce post-harvest food losses

Roots and Tuber Crops

The principal roots and tubers are derived from nine species. They are all of high Moisture content, and their parenchyma cells are packed with starch grains. These materials show variable degrees of inherent keeping life, from some species of yam, like Dioscorea alata which remains dormant for 3-4 months to cassava in which there is no natural dormancy. When such material is stored in the fresh state, respiration and transpiration continue with inevitable losses of water and dry matter. Prolonged high levels of transpiration result in a change of texture affecting quality and weight loss.

The main root/tuber crops are cassava (Manihot esculenta Crantz), the yam (Dioscorea spp.), the potato (Solanum spp.), the sweet potato (Ipomoea batatas L. (Lam) and the edible aroids (Colocasia spp. and Xanthosama sagittifolium). With the yam, the potato, and the aroids many different species are used in the cultivated complex. Differences between these are usually sufficient to merit separate treatment. Tare (Colocasia esculenta var esculenta) illustrates the system of maintenance of continuous food supply well. This species which has no seasonal growth constraint is planted in moist situations conducive to year-round growth.

By systematic replanting whenever material is harvested, continuous supplies are assured. A similar absence of seasonality in cassava (Manihot esculenta) together with its ability to produce mature roots at various time intervals after planting permits year-round harvesting in most ecological systems. By opting for continuous food supply the opportunity for post-production losses is restricted to the food preparation phase. The same benefits accrue from the use of crop scheduling. Maintenance of food supplies ever restricted periods through storage is an essential feature of yam-based food supply systems.

The success with yam storage lies in recognition of the limits to the potential for storage and in producing an environment conducive to keeping the material in sound condition. Perceptive farmers like those in Nigeria and the Ivory Coast often construct different stores to meet the separate requirements of D. alaṭa and the different culivars in the D. rotundata/cavenesie complex. These systems work well within the normal period of yam tuber dormancy.

Considerable quantities of roots and tubers are transformed into more durable products by drying, fermentation and commination in different combinations and sequences producing a variety of materials each with distinctive characteristics. Other benefits that often accrue from transformation include the removal of toxic substances naturally present and the convenient-to-use nature of the transformed product. Although the ultimate keeping potential is constrained by environmental conditions, transformations have good prospects for long-form maintenance of root/tuber crop supply because they convert a perishable commodity into a form with stability similar to that of the cereal grains.

A feature of much of the indigenous processing technology is its high labour requirement. Because of this, food supply strategies based on transformation tend to be restricted to economic environments where labour is available. As a result root crop transformation is widely practiced among subsistence economies whore labour inputs have zero or low cash value. Attempts are being made to establish root/tuber crop transformation commercially, but success has been variable. Radical changes in the entire system of production may be required, as has been suggested with yam (D. rotundata/ cavonenais) in Nigeria for manufacture of instant pounded yam.

Factors Affecting the Post-harvest System

Technologies

The major technologies for reducing losses in horticultural products are listed below followed by a statement of probable environmental effects from the named procedure.

Gentle handling

Because of their soft texture all horticultural products should be handled gently to minimise bruising and breaking of the skin. Bruising renders the product unsaleable to most people although it usually has minor effect upon the nutritional value. The skin of horticultural products is an effective barrier to most of the opportunistic bacteria and fungi that cause rotting of the tissues. Breaking of the skin also stimulates physiological deterioration and dehydration. Careful digging and movement of roots and tubers significantly reduces postharvest losses.

Careful handling of fruits and vegetables to minimise bruising and breaking of the skin likewise is a well-known method of reducing postharvest losses as is the provision of adequate shipping containers to protect the produce from bruising and puncturing of the skin. Reducing the number of times the commodity is handled reduces the extent of mechanical damage.

There are no adverse environmental effects to this technology. Thus careful digging, harvesting and handling, and appropriate packaging end transportation are environmentally count methods for reducing losses. Also, since damaged skin is the major entry point for fungal infections, some of which produce mycotoxins, gentle handling can improve the safety of the produce.

Temperature control

It is well known that cooling horticultural produce extends

storage life by reducing the rate of physiological change and retarding the growth of spoilage fungi and bacteria. Cooling is the foundation of quality protection. There are several ways of reducing the storage temperature of horticultural crops:

Cooling technique	Environmental effect
a) Keep out of direct rays method with minimal	This is an easy low-cost of sun. effect on the environment. Almost all societies can provide shade at low economic or environmental cost.
b) Use natural cooling, e.g., harvest during the cool early morning hours, open stores for ventilation during the cool of the night, utilise the cool temperature of high altitude or a natural source of cold water when available.	Minimal environmental costs.
c) Evaporative cooling obtained by drawing dry air over a moist	Minimal environmental and economic costs. Restricted to areas of low humidity and low-cost water.
d) Mechanical refrigeration.	Energy coats and economic costs are relatively high but give most positive control of temperature. Generated heat is dumped into the environment.

e) Cool promptly after harvest. | High energy coat.

Since every degree reduction from ambient temperature increases storage life every form of cooling is beneficial even if it is not optimum cooling, i.e. simple low-cost cooling or refreshing the product is better than no cooling at all. The optimum storage temperature for most temperate horticultural crops is close to 0°C. If they are cooled slightly below this temperature they freeze and suffer from "freezing injury" and spoil quickly. Most tropical horticulture crops however can be injured oven at temperatures above freezing point. This is called "chilling injury" and causes rapid deterioration in quality.

The optimum storage temperature for most tropical horticulture crops is between 7 to 10°C; for yams and bananas it is about 15°C. Although refrigerated storage is not often appropriate for some commodities such as yams, it should be considered an important element in the temperature management of a wide range of perishable crops because it gives the most positive and direct control of temperature. The popularity of refrigerated storage in some countries has suffered setbacks duo to occasional poor design of units and bad management. This has sometimes resulted in the impression that refrigerated storage is costly and unsuited for use in developing countries which is not necessary the case.

Good cosign and proper management are essential ingredients in considering the introduction of refrigeration techniques as are the supporting infrastructures available within the post-harvest system. When studied on a case-by-case basis it seems likely that refrigerated storage will continue to find many successful applications to the needs of developing countries. There are tour basic principles which must be correctly applied for successful refrigeration of perishable crops:

i. *Select only healthy products*: Refrigeration does not destroy pathogens responsible for product deterioration but only slows their activity; it does not improve product quality but only maintains it. A damaged product will deteriorate more quickly than a healthy one even in refrigerated storage, hence it is pointless to submit poor quality produce to refrigeration. In addition storage under refrigeration increases the coat to the product. The storage therefore of low grade diseased produce frequently cannot achieve an adequate economic return.

ii. *Timely cooling*: Since refrigeration slows the development of microorganisms and physiological changes responsible for deterioration of perishable crops, it is obvious that cooling should be applied as soon as possible after harvest. The technique of pre-cooling was developed to fill this need by cooling produce soon after harvesting down to a temperature appropriate to that product.

iii. *Adhere closely to optimal conditions for temperature and relative humidity*: It is well known that refrigeration provides maximum storage life it these two parameters are correctly adhered to. This fact is especially important for tropical fruits and vegetables because their optimum storage temperatures vary considerably between varieties and even between producing areas. One of the main roles of research centres in tropical countries should be to define optimum storage conditions for commodities grown under tropical conditions. There is a need to evaluate the limitations of storage of these commodities under a range of temperature conditions and to consider the implications and problems of product compatibility under conditions of mixed commodity storage.

iv. *Uninterrupted cooling*: Refrigeration should be applied from the point of harvest through to the point of consumption where maximum post-harvest life with high product quality is justified. This is the well known concept of the "cold chain".

Humidity

High humidity retards wilting and maintains the product in better condition. Most horticultural products store best in an atmosphere that has a relative humidity of 90%. Providing humidity has little environmental cost.

Waxing

Waxing the surface of horticultural products is a treatment used on a number of commodities including citrus fruits, apples, rutabagas and cucumbers. It retards the rate of moisture loss, and maintains turgor and plumpness and may modify the internal atmosphere of the commodity, and is performed primarily for its cosmetic effect; the wax imparts a gloss to the skin and gives the produce a more shiny appearance than the unwaxed commodity. Sometimes antiwaxing is a techniques that could probably be used more widely in developing countries with advantage.

In some countries indigenous waxes may be suitable for this purpose. For example, experiments in Colombia have shown that waxing of cassava can extend the storage life from 2 to 3 days up to about 30 days by preventing discolouration in the vascular tissue. Work in India has also demonstrated the efficacy of indigenously produced wax emulsion formulations in extending the storage life of different fruits and vegetables.

The waxes and wax formulations that are used in the U.S. are approved by the Food and Drug Administration and are kept under continuous review. Most of the ingredients in the wax mixtures are classified "Generally Recognised As Safe" (GRAS). In most cases the skin is removed and discarded before consumption in which case the wax is not ingested and should cause no special problems. However, problems might arise if unregistered formulations are used, or if the skin is eaten by humans or fed to animals.

Controlled atmosphere storage

Controlled atmosphere storage consists of placing a commodity is a gas-tight refrigerated chamber and allowing the natural respiration of the fruit to decrease the oxygen and increase the carbon dioxide content of the atmosphere in the chamber. Typically, for storage of apples the oxygen content is lowered to about 3% and carbon dioxide is allowed to increase to 1 to 5%. This atmosphere can extend the storage life of apples by several months and allows fresh apples to be marketed every month of the year. This technology requires expensive storage chambers and close supervision of the composition of the atmosphere and is unsuited for widespread use in less developed countries. Some roots and tubers are stored in pits in the ground, known as "clamp storage". Well designed clamps tend to change the atmosphere to some extent by reducing oxygen and increasing the carbon dioxide content.

Modified atmosphere storage would probably be effective for a limited number of commodities in developing countries especially if coupled with low temperature storage. Short pre-storage exposure to high carbon dioxide and low oxygen atmosphere of vegetables can extend the storage life of commodities even at ambient temperature. Since this technology only manipulates the proportions of asses that are naturally present in the air there should be no adverse environmental effect.

The now technology of hypobaric storage is emerging which maintains reduced pressure in the refrigerated storage chamber by means of vacuum pumps. In this system the commodity is placed in a flowing stream of highly humidified air which is maintained at a reduced pressure and controlled temperature. Under these conditions, Bases released by the commodity that limits its storage life, are flushed away. Reports indicate that the storage life of certain fruits and vegetables is extended substantially by this procedure. The economic feasibility of this type of controlled atmosphere storage is presently being tested. This is an energy-intensive and capital-intensive technology and is perhaps unsuited for less developed countries. The major environmental effect is the high energy coat.

Field factors

Maturity at time of harvest is an important factor in the keeping quality of horticultural products. Commodities that are harvested in an immature state not only have poor eating quality but may tend to shrivel in storage and be more susceptible to storage disorders. When picked too mature the commodity is soft or fibrous, the flesh breaks down more quickly and it has a shorter storage life. There is an optimum time of harvest to give maximum storage life for fruits, vegetables and tubers.

The rootstocks used for establishing fruit orchards may affect [oases. For example, McDonald and Wutscher reported decay in grapefruit ranging from 3.3% to 27.7% depending on the rootstock. It is reported that the storage life of fresh cassava can be greatly extended by leaving part of the stalk attached to the tubers at harvest time. There are a number of other field factors that affect losses and these should be utilised as much as possible. Generally there are no adverse environmental effects in these operations.

Suberisation and curing

Potatoes, sweet potatoes, yams and several other roots and vegetables have the ability to heal skin wound e when held at moderately warm conditions end high humidity for several days after harvest. The self-healing of wounds, cute and bruises is known as curing. There are two steps in the curing process. First is suberisation - the production of suberin and its deposition in cell walls. The second is the formation of a cork cambium and production of cork tissue in the bruised area. The new cork tissue seals the cut or bruised areas and helps prevent the entrance of decay organisms.

The healing of injuries received in harvesting and handling prolongs the storage time and reduces the incidence and spread of decay in storage. The storage life of onions and garlic is extend by exposure to warm dry conditions for several days to dry the outside akin and prevent the ingress of spoilage organisms. This process if also known as curing although physiologically it is rather different and causes about 5% weight loss. Curing is carried out in the field when weather conditions are suitable; otherwise the product is subjected to forced circulation of warm dry air when first put into storage. This is sound environmental practice. There is little effect on the environment from curing.

Genetic control of shelf life

Each variety of a horticultural crop has a limited storage life even under optimum storage conditions. The potential storage life is partly under genetic control and can be manipulated by breeding. Plant breeders should be encouraged to include potential storage life as one criterion in their programme for breeding improved varieties of roots, tubers, fruits and vegetables. This is particularly needed with the breeding programmes in tropical climates where refrigerated storage capacity is in short supply. This should be a high priority method for reducing losses in horticultural products.

Farmers should be encouraged to grow varieties that hare long storage life. The U.S. Food and Drug Administration established regulations for release of new varieties of edible plans a when it wee discovered that a new potato variety that wee released in 1969 had an unusually high content of the toxic glycoalkaloids that are naturally present in potatoes. The FDA regulations apply to any plant material that provides more than 2% of the U.S. diet. The regulations require that plant breeders must establish two point a before releasing a new variety:

(i) that the content of the major nutrients is no lower than the average found in existing varieties of that commodity and

(ii) that toxic substances naturally present in the commodity are no higher than normal for existing varieties.

Time Shortage

In developing countries a considerable amount of produce is wasted because of poor transportation systems and poor marketing procedures. Much produce is spoiled because it is stored beyond its inherent shelf life before marketing is completed. Improving transportation and marketing facilities, spreading the harvest season by growing varieties that mature at different times, and staggering the planting dates of annuals and reducing the number of steps between producer and consumer are methods that can be used to shorten the time between harvest and consumption.

Processing

Considerable quantities of fruit a and vegetables are

processed by dehydration, canning and freezing in developed countries. In developing countries. In developing countries small amounts of these commodities are processed for local consumption although large volumes of some commodities are processed for export (e.g., canned pineapple). Canning and freezing require a high capital cost, high energy costs and expensive packaging and are unsuited for widespread use in less developed countries. Dehydration or sun drying is the simplest and lowest cost method of preservation and should be more widely used in developing countries because it converts a perishable commodity into a stable item with long storage life. Some excellent quality dehydrated products can be made from roots and tubers; this kind of processing should be encouraged.

Occupational hazards in the fruit and vegetable processing industry are the normal hazards associated with machinery, for which adequate safety measures are well developed. The National Institute for Occupational Safety and Health in the U.S. (NIOSH) have no complaints safety hazards in processing plants that handle horticultural products. The U.S. Occupational Safety and Health Administration (OSHA) have no regulations specific to the fruit and vegetable processing industry other than the board guidelines that apply to industry in general. The fruit and vegetable processing industry is not on the list of occupational groups in which excess cancer incidence is reported by the U.S. National Cancer Institute.

There are occupational risks to some workers with specific horticultural products. For example, Barber and Husting report isolated cases of contact dermatitis among workers handling raw fruits and vegetables, including carrots, asparagus, mangoes, cashew fruits and nuts, and some citrus fruits. Fruit and vegetable handlers may also suffer contact dermatitis due to sensitivity to specific insecticides and fungicides. Indirect effects of handling fruits and vegetables include chapping and moniliasis from exposure to moisture, photosensitisation dermatitis from sunlight, and parasitism from mites. Products that cause photosensitisation include fig, rue, lime, bergamot, paranips, parsley, carrots, fennel, dill and pink rot celery. Raw pineapple fruits contain the proteolytic enzyme bromelain which causes skin irritations to workers in pineapple processing plants. This problem is overcome by supplying workers who handle cut fruit with rubber gloves.

Heat treatment

Some of the organisms that cause rotting are inhibited or killed at elevated temperatures that are below the injury threshold of the product. For example, hot water dipping of mangoes at about 50°C for a few minutes kills many pathogens without adversely affecting the quality of mango. Heat treatment is however not a desirable procedure for most fruits and vegetables. When applicable, very rigid temperature controls are needed. There is little adverse environmental effect from heat treatment. Small amounts of heat are dumped into the environment.

Sanitation

All handling, storage, cleaning and washing equipment for horticultural products should be kept in a sanitary condition in order to minimise the risk of spreading infection. Diseased or damaged units should be sorted out and properly disposed of because their presence promotes the growth of fungi and bacteria. Insects infesting cull piles may fly to good produce and introduce pathogenic organisms and increase losses. Wash water should be changed at regular intervals before it becomes heavily contaminated with fungi and bacteria and spreads infection.

In some cases the wash water is treated with chlorine or some other chemical in order to reduce the count of viable organisms. The sanitation programme in the People's Republic of China is considered an exceptionally important element of pest control. The environmental effects of good sanitation practice are minimal.

Uses of Chemicals

A number of chemicals may be applied to horticultural products in order to obtain a desirable post-harvest effect. Most of these are applied after harvest, but a few are applied in the field in order to obtain a specific post-harvest response. For example, the sprouting of onions in storage can be delayed by spraying the onions with maleic hydrazide (MH) in the field while the tops are still green. Chemicals used pre harvest whose scale propose is to achieve a post-harvest effect should be included in the list of post-harvest chemical treatments. Post-harvest chemicals are classified into groups below. Many of these are not used commercially and are of research interest only:

a. Fungicides which prevent or delay the appearance of rota and molds in the product. Examples are, sodium orthophenylphenate (SOPP), benomyl, thiabendazole (TBZ), sodium hypochlorite, and sulphur dioxide (SO_2). Methyl formate (Erinol), ethyl formate and (in some countries) ethylene oxide are frequently applied to dried fruits to kill infestations of insects and molds. Sulphur dioxide and benzoic acid are frequently, and propionic acid, ascorbic acid or sorbic acid sometimes, added to processed fruit products, especially juices, to inhibit the growth of yeasts and molds.

b. Chemicals that delay ripening or senescence. Examples are: the kinins and kinetins that delay chlorophyll degradation and senescence in leafy vegetables, gibberellins that retard the ripening of tomatoes and hold citrus fruits on the tree beyond normal maturity, and auxins that delay physiochemical deterioration of oranges and green beans.

c. Growth retardants that inhibit sprouting and growth. Examples are maleic hydrazide which is applied pre-harvest and inhibits sprouting in a number of stored commodities, e.g., onions and potatoes. A number of chemicals are applied post-harvest to potatoes to control sprouting, for example, CIPC, TCNB and MENA. Daminozide (Alar) give a increased fruit firmness, bettor colour and early maturation in apples.

d. Chemicals that hasten ripening and senescence. Examples are ethylene and compounds such as Ephephon that release ethylene, abscisin, ascorbic acid, b hydroxyethyl hydrazine (BOH), acetylene and substances that release acetylene such as calcium carbide, and certain alcohols and fatty acids.

e. Chemicals that may hasten or delay ripening and senescence depending on the dose and the commodity on which they are used. Examples are 2, 4-D; 2,4, 5-T; indoleacetic acid (IAA) and naphthalene acetic acid (NAA).

f. Metabolic inhibitors that block certain biochemical reactions that normally occur. Examples are cycloheximide, actinomycin D, vitamin K, maleic acid, ethylene oxide, and carbon monoxide.

g. Ethylene absorbants. These delay ripening and senescence because they remove the ethylene produced by the fruit. They are usually placed in clove proximity to the commodity and leave no residue on it. An example is potassium permanganate-impregnated alumina or vermiculite (fur fir).

h. Fumigants to control insects or sometimes molds. Ethylene dibromide and methyl bromide are the most commonly used fumigants.

i. Colouring. The use of artificial colours is sometimes permitted in order to improve the appearance of a fruit. For example, fresh grange a from Florida may have artificial colour added to the akin for cosmetic purposes. Since most people do not eat orange skins other than for marmalade it is considered to be a harmless addition. In warm climates ethylene is used to degreen lemons, oranges and tangerines imparting a brighter colour to the skin. Ethylene is a naturally occurring metabolite of ripening fruits.

j. Food additives. A number of compounds are permitted to be added to processed horticultural products for preservative or functional effect. The major preservatives are sulphur dioxide, benzoic acid or benzoates, and sorbic acid or sorbates. Functional additives include antioxidants, colouring, flavouring, thickeners, emulsifiers, etc. The use of food additives in the U.S.A. is regulated by the Food and Drug Administration (FDA).

Other countries have an equivalent government agency to regulate the use of additives. At the international level the Joint FAO/WHO Expert Committee on Food Additives (JECFA) formulates general principles governing the use of food additives and makes recommendations regarding their examination and control. Food additives will not be discussed further because they are only used in processing and formulating horticultural product a and are not applied to raw horticultural products. One exception is sulphur dioxide which is used to fumigate fresh grapes in cold storage in order to control growth of yeasts and molds.

There are two important differences between the use of chemicals in the field and the use of post-harvest chemicals:

i. Smaller quantities of post-harvest chemicals are used. For example, the normal dose of CIPC for controlling sprouting of potatoes is about 30 grams per ton and the normal dose of ethylene dibromide for fumigation of fruit a and vegetables is about 30 grams per ton. These levels contrast with the use of field chemicals where doses of one to several kg. per hectare are commonly used.

ii. The chemicals are not broadcast over the field but are applied in the confined apace of the storage chamber.

It is impossible to obtain figures for the quantities of post-harvest chemicals that are used because this is considered proprietary information by the companies that manufacture and formulate them. However, all post-harvest chemicals are classed as "minor use" by the U.S. Environmental Protection Agency because the quantities used are relatively small. In the U.S. a company must produce experimental evidence of the toxicity, safety, and usefulness of a new agricultural chemicals before it can be registered for use as an agricultural chemical. Each use must be cleared through registration for every commodity to which it is applied.

Most of these are inert ingredients that do not affect the pest but do improve the functional properties of the pesticide formulation. Examples of these chemicals are, surfactants, solvents, diluents, synergists, preservatives, stabilisers, antioxidants, thickeners, emulsifiers, and antifoam agents. Most of these substance are Generally Recognised As Safe (GRAS) by the U.S. Food and Drug Administration. The use of post-harvest chemicals in the U.S. is strictly controlled and monitored. The chemical suppliers usually described in detail on the label and/or in supplementary literature exactly how, when, and how much of the chemical is to be used. This is backed up by the agricultural extension service of each state which keeps in close contact with the farmer.

Most states have a cadre of inspectors who regularly draw samples of horticultural products and submit them to a central analytical laboratory to assay for chemical residues. Few violation of the regulations are detected, and most samples tested are found to be well below the tolerance. Most other developed countries maintain close supervision and control of the use of post-harvest chemicals on horticultural products. The situation may be quite different in the less developed countries where governments usually do not have the expertise or back-up analytical laboratories to monitor adequately the use of post-harvest chemicals on perishable crops.

It is difficult to obtain information on this topic, but from a general knowledge of how governments in LDCs operate, it appears to be a matter that deserves investigation. Presumably, the pesticide tolerances for the major export crops (e.g., fresh bananas) are effectively monitored by the large corporations who operate this trade and by the developed countries that import these commodities. One cannot be so sanguine about the horticultural crops that are indigenously produced and consumed in the less developed countries. The U.S. Cancer Institute has prepared a list of 26 chemicals or industrial processes associated with cancer induction in man.

The whole issue of harmful chemicals in the environment is complex and not very clear at the present time. However, we can make two reasonable assupmtions about chemicals added to horticultural products:

1. The developed countries have the expertise to engage in the debate on the harmfulness of chemicals, to evaluate the risks and benefits of their use, to enact new legislation controlling their use as new knowledge becomes available and to establish the inspection-analysis-prosecution machinery to ensure that the legislative intent is carried out. It is reasonable

to assume that any proven grave rick from the use of chemicals will soon be brought under control in the developed countries.

2. Most of the less developed countries have little of the kind of expertise listed above. There is a real risk that chemicals will be improperly used and their improper use will not be brought under control, with the potential to cause harm to human health and the environment. Therefore, the use of post-harvest chemicals in a given country should be discouraged until adequate inspection services analytical laboratories have been established to ensure that these chemicals are used safely..

Misuse of certain post-harvest chemicals may lead to seriuos environmental harm. As far as can be determined little is known about the ultimate effects of post-harvest chemicals on the environment when they are correctly used. It seems to be generally assumed that since the compounds are used in small quantities in confined areas, and since most of then decompose into nonactive substances there are no adverse environmental effects. Although this is a reasonable assumption there is little concrete evidence either for or against this widely held opinion.

Pests

The major causes of lose in perishable produce after harvest are certain pathogenic fungi and bacteria. Viruses and nematodes play a minor role in postharvest losses; rodents and insects are also generally of lesser importance in contrast to the significant damage they cause in food grains. In addition to the direct lone in quantity of food resulting from microbial infections, a partial loss results because of effect on appearance and/or quality resulting from disfiguring surface infections of fruit, root, and tuber crops. Other secondary adverse effects may include a decline in shelf life, possible contamination with mycotoxins, acceleration of ripening because of release of ethylene in pathogenesis by certain fungi, and in some instances deterioration of canned fruit crops because of the presence of heat-resistant hydrolytic enzymes formed by decay fungi in fruit tissues prior to the canning process.

The loss in the post-harvest period may originate from infections that were initiated by fungi during the growing season well in advance of harvest. Much of this pre-harvest infection involves a group of fungi that are capable of infecting healthy developing fruits either by direct penetration, e.g., anthracnose deseases caused by species of colletctrichum or by invasion via natural openings much as lenticels or stomates or through breaks in the tissue at the points of attachment of fruits to the plant. In many cases the infection process may be incomplete. Thus, sub-cuticular mycelium may be formed which remains in a latent stage until the post-harvest period when changes in susceptibility may occur and the pathogen mycelium may ramify through the tissue.

Many of the fungi and bacteria nvolved in decay problems associated with the post-harvest period may be considered as opportunistic pathogens. They are usually incapable of penetrating unijured tissue or aggressively attacking vigorous healthy plants during their active growth period. However, they do have the ability to parasities fleshy plant organs when tissues are bruised, injured by insects, or otherwise placed under environmental stress. In many cases tissues are invaded by a succession of organisms which may interact in a synergistic manner.

Each species of root, tuber or fruit is affected by specific groups of fungal or bacterial pathogens. It is important that broad non-specific designations of these organisms be avoided (i.e., molds or rot organisms). The species of fungi or bacteria associated with specific decay problems should be properly identified. For example, it has been clearly shown that species of Knisopus differ markedly in their susceptibility to specific fungicides. The main factors affecting disease development are:

1. Most susceptibility
2. maturity
3. wounds and wound healing
4. temperature of the commodity
5. relative humidity (especially in storage)
6. packaging
7. handling in general
8. concentration of inoculum

The basic methods of control involve three different approaches 1) prevention of infection 2) elimination of incipient or latent infections, and 3) prevention of spread of the pathogen in the host tissue. Losses due to microorganisms may be reduced by:

1. refrigeration
2. improved handling procedures
3. pre- and post-harvest chemical control

Although every effort needs to be made to minimise or reduce dependence on chemicals, in many cases no viable alternatives exist. A wide range of fungicides are now available that are effective and safe to use. Incipient infections can be eliminated or reduced by application of certain fungicides such as benomyl which have the capacity of diffusing into host tissue and killing the pathogen in situ. Certain fungal pathogens initiate infections of fruits and vegetables during the growing season and these can be controlled best by timely application of fungicides prior to harvest. It is essential to ensure that chemical controls are used properly.

Major Root/Tuber Crops

Cassava (Manihot esculenta Crantz)

Known variously as manioc, yuca, tapioca and mandioca, the edible portion of cassava is a starchy root which matures in harvestable state in 8-24 months according to cultivar and climate. In addition to the roots, the leaves also have potential for use as food as for instance in pert e of Central and West Africa. Cassava cultivators are often grouped into "sweet" with relatively low contorts of cyanogenic glucosides and "bitter" with high cyanogenic glucoside content although many intermediate forms exist. The "sweet" types may be eaten raw or lightly boiled without harm while the "bitter" forma require processing to remove cyanogenic glucosides.

The cassava roots have no natural dormancy and are highly susceptible to deterioration. The nature of this deterioration is as yet not fully understood, although two factors, one an indigenous physiological one and the other associated with microbiological infection, have been identified. The physiological deterioration of which the symptoms are tissue discolouration, can commence as soon as 24 hours after harvest. Acting independently of, but usually following, the physiological changes, microorganisms, both fungi and bacteria develop in the flesh of the roots causing additional damage and loss: this microbial deterioration usually develops only 5-7 days after harvest. This rapid post-harvest deterioration of cassava roots places serious constraints on their use with fresh produce trade and on the holding of buffer stocks for large-scale processing.

The short storage life of fresh cassava imposes constraints on its distribution and use. As a result, the choice of market for which production is intended becomes influenced by location. The range over which production for the fresh market extends is therefore a function of the distance from their markets and the efficiency of transport. The time of delivery rather than distance per se is the important limiting factor. In the vicinity of Bogota, Colombia, up to an 8-hour interval from harvest to delivery is acceptable. With the current status of road a and vehicles this permits production within a 300 km radius. With more effective preservation or a faster delivery system, this range could be extended. It is claimed that the market for fresh cassava in Colombia is so structured that as quality deteriorates, the price is lowered, thereby enabling roots of decreasing quality to be bought by successively poorer income groups.

Techniques to extend the fresh life of cassava roots have been independently developed in several countries. At the Instituto de Investigaciones Tecnologicas in Colombia deterioration has been delayed by coating tubers with a film of paraffin wax. At CIAT fresh tubers were shown to have improved keeping characteristics when plant stoma are out and removed but tubers are loft unharvested for up to 14 days. Tubers so treated remained sound for as long as 30 days. This technique has much to commend it, but has not yet been tested on a large scale. Other practices developed in joint CIAT/TPI projects include storage in earth silos' and storage of roots in plastic bags and in absorbent packs. These methods appear to have some utility in small and medium scale production but have not yet been proved in large-scale commercial operations.

None of those techniques have as yet attained general application. The transformation of cassava to more durable forms, frequently in forms more convenient to use than fresh roots, is partly associated with the detoxification process for the removal of hydrocyanic acid. From the crop loss point of view these products are important since they are frequently more able to avoid the rapid deterioration of the fresh tubers. Cassava products may be classified in a variety of ways, but for simplicity these may be regarded as food starch ant whole root products. These latter may be fermented or unfermented and may either be "flour" or "gari" types.

Cassava starch is extracted from roots after pealing, washing, reaping, squeesing settling and drying. Two types of starch are produced, sour or fermented and sweet or unfermented. Sweet starch is dried immediately following extraction while sour stanch is stored for 3-4 weeks when some fermentation involving lactobacilli and yeasts occurs. Sour starch is favoured for baking and commands a better price than sweet starch at least in some countries. This latter has similar conformation to corn starch with which it competes disadvantageously. Gari, the most commonly used form of cassava in West Africa, accounts for some 70% of the entire cassava production in Nigeria.

It is estimated that between 4 and 5 million tone of roots are used each year for this purpose. Although there are close parallels between the production of gari and of Brazilian "farinha de mandioca", the West African variant has some distinctive features. The effect of the longer fermentation affects the taste by the production of lactic acid, as wall as well as reducing the content of free hydrocyanic acid. Production at levels of indigenous technology is highly labour-intensive, but this constraint is being progressively removed through a variety of innovations aimed at increasing labour productivity. The hygroscopic nature of gari is a major constraint to its keeping quality.

In a humid atmosphere it can absorb sufficient moisture to make it vulnerable to the growth of fungal organisms. The recent practice of packaging gari dried to a safe moisture contort of around 12% in sealed polyethylene bags enables the product to be kept in good condition for extended periods. Farinha de mandioca is the Brazilian product from which gari derives. The stages of preparation are similar to those for gari, the roots being washed, peeled, grated, pressed, toasted and classified by size. Unlike gari, the mashed or grated mass is not permitted to ferment. The heating dries the product and allows some gelatinisation and dextrinisation of the starch as also occurs in gari making. Other durable cassava products include cassava flour, dried chips and cossettes from which flour can subsequently be prepared by milling or grinding.

Tam (Dioscorea Spp.)

With yams, storage life is determined by dormancy since attempts to store yams after sprouting has commenced are impractical. During the period of dormancy the metabolic processes continue at reduced rate thereby keeping dry netter losses at relatively low levels. The various yam species have different periods of dormancy with 3-6 months for D. rotundata and D. alata and 1-2 months with D. cavenensis or D. trifida. Pathological factors are also of great importance in causing postharvest losses in yams. Traditionally, in West Africa, yams are stored in "yam barns" where good but uncontrolled conditions for storage exist by building the barns in the deepest available shade. This prevents heating by the nun and also helps to maintain high humidity.

Several other indigenous yam storage systems are used but moat of these are less satisfactory than the "barn" or "claie verticale". It is generally believed that these systems are well adept d to the yam varieties and to the climatic conditions prevailing in West Africa and offer little scope for improvement at current levels of technology. Of the methods used to extend storage only the use of ionising radiation has been an unreserved technical success. The practical application of this technique is not yet generally available. Reduced temperature storage is limited by possibility of chilling damage around 10-12 C and by the growth of fungal organisms at temperatures around 14° C.

Treatment with chemicals used to suppress sprouting in other crops has had little success but active research is being pursued by TPI. Controlled gas storage has not thoroughly been investigated. Transformation of yams to more durable forma is possible but these methods have not yet met with much commercial success. Traditional products such as "amala" are variable. This may be related to the intrusion of undesirable dark colours into the final product from the use of a mixture of varieties and the failure to deactivate the polyphenol oxidases during processing rather than to any inherent basic defect of the process.

The attempts of commercial organisations to produce an instant pounded yam failed more as a result of nontechnical problem such as the logistics of raw material supply than to technical defects in the process. New processes recently developed at the University of Ife, Nigeria, may provide the required answer. A yam flake product developed to pilot project level in the West Indies has not continued long in production at a commercial level.

Potato (Solanum Spp.)

The potato, Solanum tuberosum, originally of highland tropical origin, but developed mainly as a temperate crop, is now being developed as a lowland tropical crop. Selections tolerant to high temperature and short days are being made in several parts of the tropics. At higher elevations in the tropics it is already established as a crop of some importance. Several products are made from potatoes including chuno, papa seca, tungush and potato starch. Ecological zone and economic circumstances result in different comparative advantages between choice of product.

The most common product at low elevation is papa seca, a parboiled dried product which requires, fuel for boiling but which utilised what would otherwise be a waste product. This restricts production to the lower zones where fuel supplies are more assured. Hybrid or native varieties are used for papa seca since this process would not be sufficient to remove the glycoalkaloids present in the bitter varieties. Processing into chuno or tongosh which eliminates the bitter principle requires no cooking.

The process however requires heavy night frosts and open areas on which to expose the potatoes to natural freese drying. These conditions are all readily met in the high blue. Locally produced potato starch competes with commercially produced corn starch. It is clear from this example that changes due to the technological superiority of one process over another is unlikely since a different technique is adapted to meet different complexes of factors.

Sweet potato (Ipomoea batatas (L.)

Sweet potato is a crop of considerable unrealised potential. The roots are widely used as a carbohydrate food and recently attention has also been drawn to the nutritional value of its leaves as food. The ability of the roots to become suberised at high temperature and humidity in the classical curing process, gives the crop potential for extended storage. Storage is particularly successful when kept at relatively cool temperatures, although, it held below 12°C, chilling damage can occur.

The most successful storage occurs in the southern states of America and in Japan, and successful storage for extended periods has still to be achieved in the Tropics. There is interest in transforming the roots to more durable forms. In Trinidad, at the University of the West Indies, a sweet-potato-based flour was shown to have utility in wheat flour substitution in bakery products. There is interest there and elsewhere in the preparation of breakfast foods from sweet potato tubers. The manufacture of dried sweet potato flakes is already an established industry in the United States.

Taro (Colocasia esculenta)

The edible corms of these crops are normally eaten fresh after boiling or baking. Estimates in Fiji showed high peeling and trimming losses during preparation as food. The fresh corms to not normally remain sound for long periods, and signs of deterioration become apparent a few days after harvest. Observations in Fiji indicated that packaging in plastic bags enhances keeping quality. In the South Pacific it is traditional to keep taro by transforming it into a fermented product of which "Poi" is the Hawaiian variant.

Fruits and Vegetables

Many post-harvest losses are direct results of factors before harvest. Fruit a and vegetables that are infected with pests and diseases, inappropriately irrigated and fertilised, or generally of poor quality before harvesting, can never be improved by post-harvest treatments. Very often the rate of commodity loss is faster if the quality at harvest is below standard. Thus, the processes in the attainment and maintenance of quality from production, harvesting, handling and marketing must be considered a unified system. The success of preserving the harvest-fresh quality of produce demands control of each step in the system, depends on the previous stop and therefore is a chain of interdependent activities. The small size and isolation of many vegetable farms make it logical to employ manual harvest labour which is often relatively cheap and non-organised.

Traditional methods of harvesting are still employed using a minimum of mechanical aide. Many of the farmers may know their market but do not analyse it and plant without much concern for it. Price considerations are usually given more weight than the quality of the produce. Early harvesting of carrots, chayote, snap beans, squash and bottle gourd give better quality, but lower yield. On the other hand, vegetables may be harvested before reaching prime quality if the prevailing market price is high duo to the scarcity of the product. Knowledge of maturity indices is often inadequate in moat instances visual indices being used. Therefore, more experienced farmers can deliver better quality products than those with less experience. The deterioration of a product starts during the harvesting operation.

The more carefully a product is handled, the slower the deterioration process during subsequent handling operations. However, the farmers mar be unaware or indifferent to the condition of the product after harvest, and harvesting procedures may thus be rather careless. The only constraint is to avoid external injury. Until farmers are convinced that careful handling will increase profits, it will be difficult to persuade them otherwise. In many countries contract buying is practiced where the contractor takes charge of the harvesting and may exert strict supervision of the operation.

In meet of the developing countries the world, harvesting methods are very simple. Picking poles, to which a hook or cutting knife is attached are generally used. Fruit blemishes and injury are usually unavoidable. Quite often the fruits fall to the ground becoming subject to field infestation. Planning, preparation and organisation in the harvesting operations are necessary. Suitable harvesting tools, hand gloves, containers and supplies are needed by the harvesters. The grower should exert strict supervision on the harvesting operation. Grading is a thing that most farmers are loath to do until they are convinced it will bring them added revenues or increases the acceptance of their products.

A farmer may separate different varieties if they are distinguishable, but consumers in the tropical countries usually are more price conscious than quality conscious. That is not to say that they are unmindful of the quality, but if good quality products are too highly priced, they are often willing to settle for a poorer quality product. With the variety of vegetables offered for sale, shore is usually no difficulty in Betting cheaper substitutes. Thus, initially, for local markets, grading should be limited to only what is necessary.

Deformed fruits and those with splits, punctures and incipient rotting should be removed. If the market is conditioned to accepting sound fruit with harmless surface blemishes (e.g., russet, windscarring, or mite injury), it is well not to attempt to change the situation for the time being. The following are likely to occur in the absence of proper sorting and grading procedures:

i. Presence of rotten items which contaminate food products at later handling stages.

ii. Customer deception wherein good produce on top of the containers conceals items of low quality at the bottom of the pack.

iii. Growers do not maximise their income because best prices for their product are not obtained.

Containerisation is probably, the weakest aspect in the distribution chain of fruits and vegetables. In general, packaging materials of unsuitable quality, such as large sacks, rough wooden boxes, second-hand cartons, bamboo baskets, or rattan containers are liable to cause produce to suffer from bruising, crushing and puncturing. The use of suitable containers alone will do much to maintain the quality of fruits and vegetables. Farmers often selected the cheapest and most readily available containers. For example, the use of bamboo baskets has several disadvantages. The sides are sharp and easily bruise the produce. They are too deep and without sufficient side reinforcement, thus causing the produce to be Jarred or compressed.

Moreover, handlers tend to throw rather than lift the produce gently because of its weight. The use of wooden crates could solve many of the objections to the bamboo baskets. The main objection to a wooden crate. In many developing countries wholesale markets, if any are integrated with retail markets. Ideally, wholesale markets should be separated from the retail ones. However, this is influenced by the structure of the marketing chain and the scale of the industry. Frequently wholesale markets are overcrowded, unsanitary and lack suitable facilities for display, storage, ripening, loading and unloading.

A main contributory factor to loss of leafy vegetables is trimming at wholesale markets prior to delivery at retail stands. Examination of trimmings showed that wholesalers and retailers trimmed their produce mainly because of the presence of decaying portions of leaves, due to bacterial soft rot. Thus, prevention of rotting during transport and storage would result in substantial reduction for postharvest losses of leafy vegetables. Storage of fresh fruits and vegetables prolongs their usefulness, checks market gluts, provides a wider selection of fruits and vegetables throughout the year, helps in orderly marketing and may increase the financial gain to the producer.

Adequate storage may reduce subsequent losses, but cannot overcome pre-storage losses. Adequate storage involves proper regulation of temperature, humidity, air circulation, proper stacking pattern, regular inspection, and prompt produce disposal as soon as maximum storage life has been attained. The feasibility of the construction of cold storage facilities and the interest among farmers and handlers to utilise the cold stores would again depend upon the economics parameters of the project. One of the greatest impediments to preserving quality through refrigerated storage is the consumers' strong preference for freshly harvested produce and resistance to stored produce.

Information on the storage temperature and humidity requirements of fruits and vegetables and the length of time they can be kept without decline in market value is either inadequate or unknown to those who need the information. If a farmer is persuaded to store his produce in cold storage and the market value decreases due to inadequate knowledge of the proper utilisation of cold storage, it is not only he who will become disillusioned, but his friends also will be convinced of the non-profitability of cold storage. Lack of capital may also force farmers to ignore the use of cold storage, even when available and effectively managed.

Many growers depend on almost daily sales for their income and hence may be forced to accept a lower price immediately, rather than to store their produce in the anticipation of a higher price. There is also the storage rental price which the farmer may not be willing to pay unless he is thoroughly convinced that he will not only recover his investment, but will also profit. The retailer usually disposes of produce which has been damaged by factors occurring further back in the marketing chain. At this stage, deterioration of perishables has already progressed to such an extent that street vendors have little opportunity to prevent further losses, either through storage or other preventive measures, except by estimating their potential sale for the day and buying only the amount that can be disposed of.

Miscellaneous losses are numerous. The moat important ones are:

a. Over-purchase of cheap but highly perishable fruits and vegetables leading to wastage due to inadequate storage facilities.

b. Rates of pay among "cargadores" (product haulers) usually depend on the number of containers that they can carry from one point to another. Hence, they disregard proper handling in their hurried attempts to make more trips.

c. Deterioration during storage occurs because some old stocks are intentionally kept too long in anticipation of eventual price increases;

d. Maintenance of transport, storage and other handling facilities are generally poor in developing countries resulting in a continual source of losses.

e. There is no efficient communication link between producers and wholesalers. Losses will always occur in the absence of a dependable communication system.

Following are some of the major problems of the more important commodities in the fruit and vegetable group. They are, however, indicative rather than exhaustive.

Bananas and Plantains: Harvesting is generally a one-man operation which frequently results in bruising and abrasions of fruits causing accelerated ripening and consequent decay. Latex staining is prevalent during dehanding. Bunches or hands are often piles one on top of the other without proper protection and fingers are easily detached and oftentimes wasted during transport.

This is particularly true of the very open, loose bunches of certain plantain cultivators. In container transport loosely packed hands suffer considerable damage, especially on rough roads. In transit ripening and decay are usually high, notably over long distances.

Mango: Fruits are usually harvested at the time of the day when maximum latex flow is favoured. Latex stain is allowed to dry on the peel, hence immediately reducing consumer acceptability during retail. The collapsibility of the non-rigid crates often used for transport further aggravates quality lose by compression and bruising. The inaccessibility of production area to roads causes serious delays in transport, in addition to a mixed-cargo typo of transport. Stacking in vehicles often does not provide for adequate ventilation. Loading and unloading operations are rather rough. Cold storage is not usually practised. Ripening is mainly aimed at improvement of the appearance for sales purposes and not for maintaining quality.

Papaya: Picking poles injure the fruit and there is a relatively high percentage of fruit dropping on the ground, causing breakage and bruising of ripe fruits. Peduncles are not usually trimmed hence injuring other fruit a within the pack. Rigid containers are not adequately linde, and within a single pack fruits of assorted sizes and maturity stages are often found. In bulk transport fruit e arc piled one on top of the other without any suitable padding materials. High percentage of decay, particularly anthracnose, is the main problem during ripening and in retail.

Citrus: Improper time of harvesting greatly enhances rind injury or oleocellosis. Leaving long stubs on fruit a injures other fruits within the pack. Containers used are large and overpacked, generally without sufficient ventilation. Containers are piled high with the bottom crates bearing the full weight of crates on top. Delays in transport due to poor roads often cause over-ripening or yellowing of the commodity. Poor storage conditions favour decay and physiological disorders such as chilling injury can also occur if cold stores are not wolf managed.

Grapes: Grapes are attacked by Botrytis, Cladosporium and Alternaria during storage. However, if the storage temperature is strictly maintained between 0° and 2°C, fungal attack can be reduced to a minimum Other loss factors could be berry drop, bruises, injury, water loss, and cracking of berries. Selection of unsuitable container type for packaging of grapes may also lead to heavy transit losses. Transit delays, adverse weather conditions and improper type of carriages, e.g., steel wagons, particularly during hot months may further aggravate transit losses.

Tomato: Tomato fruits are usually picked when fully ripe, and are therefore very susceptible to cracking, bruising, and consequently decay. Packaging containers often used are deep bamboo crates with insufficient aide reinforcements allowing jarring and compression during transport. Loading and unloading operations are very crude. Handlers tend to throw the pecks rather than lift them gently, on account of their weight. During retail sellers tend to pour the contents of the pack into another container, rather than transferring the fruit gently, thereby increasing bruise damage. Fruit a at the breaker stage are mixed with the fully ripe or three-quarters ripe fruits reducing the market value of the pack. Shrivelling percentage can be high since fruits are often exposed to the sun.

Onions: Insufficient grading is still existent. Spouted, injured and partly decayed bulbs are usually mixed with sound bulbs in a pack. The use of slatted wooden creates is advantageous, especially during transport. Mesh bags of 40-50 kg. capacity are also used. Sacks are thrown rather than lifted, on account of their weight. Packs are piled one on top of the other with no provision for adequate ventilation. Pre-harvest spraying with sprout inhibitors is seldom practised resulting in .serious sprouting during storage.

Cabbage/Lettuce: Improper harvesting tools contribute greatly to damage to the produce in the form of out a and abrasions. Trimming of outer leaves is usually not practised. In container transport, large crates are used (50 kg. capacity). Bruising and tearing of the leaves is of common occurrence due to the sharp edges of the containers. Containers are piled one on top of the other with the bottom crates carrying the weight of the heads above. Bulk transport likewise results in higher losses.

Peas and Beans: Factors such as the method of packing, suitability of containers, mode of transport, distance covered, number of transshipments, handling, and storage facilities in the consuming centres, all contribute to the degree of loss reported.

Bibliography

Anon, *Health risks of genetically modified foods.* (Editorial) Lancet 353:1811, 1999.

Betz, F.S., B.G. Hammond and R.L. Fuchs "Safety and Advantages of Bacillus thuringiensis-Protected Plants to Control Insect Pests". *Regulatory Toxicology and Pharmacology* 32:156-173, 2000.

Cieslak, D. "Implications of GMO's For Animal Nutrition and the Feed Industry". *61th Minnesota Nutrition Conference & Minnesota Soybean Research and Promotion Council Technical Symposium,* September 19-20, 2000, Bloomington, MN. Conference Proceedings p 72-77, 2000.

Daenicke, R., K. Aulrich and G. Flachowsky., *GMO in animal feedstuffs: Nutritional properties of Bt-maize unaffected.* Mais, September 1999, pp. 135-137.

Folmer, J.D. G.E. Erickson, C.T. Milton, T.J. Klopfenstein and J.F. Beck., Utilization of Bt corn residue and corn silage for growing beef steers. J. Animal Sci. 78 (Suppl. 2):85 Abstract 271, 2000.

Fuchs RL, Re DB, Rogers SG, et al. "Safety evaluation of glyphosate-tolerant soybeans". *In: Food safety evaluation. OECD Documents,* Paris, pp 61-70, 1996.

Hartnell, G.F. "Benefits of Biotech Crops For Livestock Feed". *Proceedings 2000 Cornell Nutrition Conference For Feed Manufacturers.* October 24-26, 2000

Horton R, *Genetically modified foods: "absurd" concern or welcome dialogue?* (Commentary) Lancet 354:1314-1315, 1999.

Padgette SR, Re DB, Barry GF et al., "New weed control opportunities: development of soybeans with a Roundup ReadyÔ gene". *In: Duke SO (ed) Herbicide-resistant crops: agricultural, environmental, economic, regulatory and technical aspects.* CRC Press, Boca Raton, pp 53-84, 1996.

_______________, Taylor NB, Nida DL et al. "The composition of Glyphosate-tolerant soybean seeds is equivalent to conventional soybeans". J Nutr 126:702-716, 1996.

Rogan GJ, Bookout JT, Duncan DR et al. "Compositional Analysis of Tubers from Insect and Virus Resistant Potato Plants", J Agric Food Chem 48: 5936- 5945, 2000.

Sauber, T." Performance of Soybean Meals Produced From Genetically Enhanced Soybeans". *61th Minnesota Nutrition Conference & Minnesota Soybean Research and Promotion Council Technical Symposium,* September 19-20, 2000.

Index